Praise for
Sowing Seeds in the Desert

"Masanobu Fukuoka ran a course on natural farming and gave our Howard lecture at Navdanya's biodiversity farm in the Doon Valley of India, and we even have a cottage named the Fukuoka hut. He was a teacher ahead of his time. Sowing seeds in the desert is what all of humanity has to learn to do, whether it is in an economic desert created by Wall Street or an ecological desert created by globalized corporate agriculture."

—**Vandana Shiva**,
founder of Navdanya Research
Foundation for Science, Technology, and Ecology

"Distilling what he has gathered from a lifetime of learning from nature, Masanobu Fukuoka offers us his gentle philosophy and a wealth of practical ideas for using natural farming to restore a damaged planet. *Sowing Seeds in the Desert* will persuade any reader that the imperiled living world is our greatest teacher, and inspire them to care for it as vigorously as Fukuoka has."

—**Toby Hemenway**,
author of *Gaia's Garden*

"From our first meeting with Fukuoka-sensei in the late 1970s at Green Gulch Farm Zen Center, he has served as a primary guide, teacher, and inspiration in the engaged practice of organic farming and Zen meditation. Now, with *Sowing Seeds in the Desert*, Fukuoka-sensei's teaching of natural farming continues to grow, sending deep roots down into the terrain of global restoration and food security for a hungry world. This wonderful book is to be celebrated and savored for its grounded, encouraging wisdom."

—**Wendy Johnson**,
author of *Gardening at the Dragon's Gate*

"This book is not a breath of fresh air, it's a howling gale from the East. It challenges us to think outside our normal, rational frames and venture into a whole new way of relating to spirituality, the Earth, and the growing of food. As I read, I was tempted to pick holes in Fukuoka's prescriptions for greening the world's deserts, but I kept coming back to the inescapable fact that he farmed his own land according to these principles over many years and produced *a lot* of food."

—**Patrick Whitefield**,
author of *The Earth Care Manual*

"Fans of Fukuoka's *The One-Straw Revolution* will be delighted by *Sowing Seeds in the Desert*, his last book. It is a rich treasure trove detailing how his own philosophy of farming evolved and how he decided to apply what he learned on his own farm in Japan to other parts of the world. His insights into the tragedies of taking Western, industrial agriculture to places like Africa to 'enrich the national economy,' and his alternative approach of working with indigenous farmers to enable them to become self-sufficient are instructive for all of us."

—**Frederick Kirschenmann**,
author of *Cultivating an Ecological Conscience:
Essays from a Farmer Philosopher*

"This book is a bombshell. Forget the gentle and retiring farmer of *The One-Straw Revolution* fame, replaced now by a flaming, world-traveling revolutionary. To achieve the kind of natural farming that can avoid worldwide collapse, Masanobu Fukuoka bluntly and fearlessly insists that we must first reject traditional ideas about God, the afterlife, accepted economic systems—especially capitalism, much of current agricultural thinking including organic farming, and even parts of science that he says are based on mistaken notions about the connection between cause and effect. Once we return to a way of life dictated by nature, not institutional religions, he says, we can apply his unorthodox farming methods to make the deserts bloom and the green fields stay lush without much expense or even labor involved. Be prepared to be mystified, irritated, shocked, and maybe even, if you persevere to the end, enlightened and encouraged by this trail-blazing book. Disagree with Fukuoka's provocative pronouncements at your own risk. Some of what he predicted in this book, originally written in Japanese in the 1990s, has already happened, especially the collapse of the Japanese economy in recent years and the spread of deserts throughout the world."

—**Gene Logsdon**,
author of *A Sanctuary of Trees*

Sowing Seeds In The Desert

Sowing Seeds In The Desert

Natural Farming, Global Restoration,
and Ultimate Food Security

Masanobu Fukuoka

Edited by Larry Korn

Chelsea Green Publishing
White River Junction, Vermont

Copyright © 2012 by The Masanobu Fukuoka Estate

Translated into English and adapted from the book originally published in Japanese in 1996 by Shou Shin Sha, Japan, as *The Ultimatum of GOD NATURE*.

All rights reserved. No part of this book may be transmitted or reproduced in any form by any means without permission in writing from the publisher.

English adaptation by Larry Korn

Unless otherwise noted, all illustrations copyright © 2012 by The Masanobu Fukuoka Estate.

Project Manager: Hillary Gregory
Developmental Editor: Makenna Goodman
Copy Editor: Laura Jorstad
Proofreader: Helen Walden
Designer: Melissa Jacobson

Printed in the United States of America
First printing April, 2012
10 9 8 7 6 5 4 3 2 1   12 13 14 15 16

Our Commitment to Green Publishing
Chelsea Green sees publishing as a tool for cultural change and ecological stewardship. We strive to align our book manufacturing practices with our editorial mission and to reduce the impact of our business enterprise in the environment. We print our books and catalogs on chlorine-free recycled paper, using vegetable-based inks whenever possible. This book may cost slightly more because it was printed on paper that contains recycled fiber, and we hope you'll agree that it's worth it. Chelsea Green is a member of the Green Press Initiative (www.greenpressinitiativet), a nonprofit coalition of publishers, manufacturers, and authors working to protect the world's endangered forests and conserve natural resources. *Sowing Seeds in the Desert* was printed on FSC®-certified paper supplied by Thomson-Shore that contains at least 30% postconsumer recycled fiber.

Library of Congress Cataloging-in-Publication Data
Fukuoka, Masanobu.
  Sowing seeds in the desert : natural farming, global restoration, and ultimate food security / Masanobu Fukuoka ; edited by Larry Korn.
      p. cm.
  "Translated into English and adapted from the book originally published in Japanese in 1996 by Shou Shin Sha, Japan, as The Ultimatum of God, Nature."
  Includes bibliographical references.
  ISBN 978-1-60358-418-0 (hardcover) — ISBN 978-1-60358-419-7 (ebook)
  1. Desert reclamation. 2. Desertification—Control. 3. Revegetation. I. Korn, Larry. II. Title.

S613.F85 2012
631.6'4--dc23
                         2012007330

Chelsea Green Publishing
85 North Main Street, Suite 120
White River Junction, VT 05001
(802) 295-6300
www.chelseagreen.com

To those who will plant seeds in the desert.

Masanobu Fukuoka
December, 1992

Contents

Introduction, xi

Editor's Notes, xxix

About the Illustrations, xxxiii

1: **The Call to Natural Farming**, 1

    My Return to Farming, 4
    Challenges During Wartime, 6
    The True Meaning of Nature, 8
    The Errors of Human Thought, 9
    No God or Buddha Will Rescue the Human Race, 13
    The Dragonfly Will Be the Messiah, 14
    A Life of Natural Culture, 15

2: **Reconsidering Human Knowledge**, 21

    The Birth of Discriminating Knowledge, 21
    Darwin's Theory of Natural Selection, 23
    Understanding True Time and Space, 25
    The Rising and Sinking of Genes, 27
    An Alternative View of Evolution, 29
    Naturally Occurring Hybrids in My Rice Fields, 31
    Abandoning What We Think We Know, 34

3: **Healing a World In Crisis**, 41

    Restoring the Earth and Its People, 42
    In Nature, There Are No Beneficial or Harmful Insects, 43

Eastern and Western Medicine, 44
The Fear of Death, 47
The Question of Spirit, 49
The Money-Sucking Octopus Economy, 50
The Illusion of the Law of Causality, 56
The Current Approach of Desertification Countermeasures, 60

4: **Global Desertification, 69**

Lessons from the Landscapes of Europe and the United States, 70
The Tragedy of Africa, 75
Sowing Seeds in an African Refugee Camp, 79

5: **Revegetating the Earth Through Natural Methods, 85**

Agricultural "Production" Is Actually Deduction, 88
Commercial Feedlots Will Destroy the Land, Cultured Fish the Sea, 90
Sowing Seeds in the Desert, 92
Creating Greenbelts, 95
The Revegetation of India, 99
Notes from an International Environmental Summit, 113

6: **Travels on the West Coast of the United States, 121**

Farmers' Markets, 124
Urban Natural Farms, 128
People Sow and Birds Sow, 129
Rice Growing in the Sacramento Valley, 134
From Organic Farming to Natural Farming, 136
Two International Conferences, 141
Japanese Cedars at the Zen Center, 145

Appendices

Appendix A: Creating a Natural Farm in Temperate and Subtropical Zones, 151

Appendix B: Making Clay Seed Pellets for Use in Revegetation, 161

Appendix C: Producing an All-Around Natural Culture Medium, 166

Introduction

MASANOBU FUKUOKA (1913–2008) was a Japanese farmer and philosopher from the island of Shikoku. His natural farming technique requires no machinery or fossil fuel, no chemicals, no prepared compost, and very little weeding. Mr. Fukuoka did not plow the soil or hold water in his rice fields all season long as farmers have done for centuries in Asia and around the world. And yet, Mr. Fukuoka got yields comparable to or higher than the most productive farms in Japan. His method created no pollution, and the fertility of his fields improved with each season.

This technique is a demonstration of Mr. Fukuoka's back-to-nature philosophy. His message is one of vision and of hope. It shows the way to a brighter future for humanity, a future where people, nature, and all other forms of life live peacefully together in abundance.

The first of Mr. Fukuoka's books to be translated into English was *The One-Straw Revolution: An Introduction to Natural Farming* (Rodale Press, 1978). As the subtitle suggests, this book was meant as an introduction both to his worldview and to the farming methods he developed in accordance with it. In the book he told the story of how he came to farm in the way that he did, with an overview of his philosophy and farming techniques. He also gave his views about such things as diet, economics, politics, and the unfortunate path humanity has chosen by separating itself from nature.

In his next English-language book, *The Natural Way of Farming* (Japan Publications, 1985), Mr. Fukuoka gave the details of how his farming techniques evolved over the years. This book was mainly practical, and, while it did not have as wide a readership as *The One-Straw Revolution*, is still well worth reading, especially for those interested in putting Mr. Fukuoka's natural farming methods into practice on their own land.

In the present book, *Sowing Seeds in the Desert*, Mr. Fukuoka gives the details of his philosophy, and introduces his plan to revegetate the deserts of the world using natural farming. It is his last work and in many respects, it is his most important.

I am not sure what I expected when I first visited Mr. Fukuoka's farm one summer day in 1973, but what I found exceeded anything I could have imagined. I had been living in Japan for several years, working on back-to-the-land communes and doing seasonal agricultural jobs when I could find them. I had heard stories about Mr. Fukuoka, always including high respect for his spiritual teachings, but no one I talked to had actually been to his farm or learned any of the details of his farming techniques, so I decided to go there and see for myself.

The rice plants in his fields were shorter than his neighbor's rice, and had a dark green, almost olive color. There were many more grains on each head, and clover and straw covered the surface of the soil. Insects were flying about everywhere; the field was not flooded, but dry. This was in stark contrast with the neighbor's fields, which were neat and tidy, straight rows of rice grown in flooded fields with no weeds and no insects of any kind. Mr. Fukuoka came to greet me and asked if I had ever seen rice like his. I told him that I had not. He said, "The reason these plants and the fields look this way is because the soil has not been plowed for more than twenty-five years."

I had heard that Mr. Fukuoka welcomed students to live and work on the farm, so I asked him if I could stay for a while. He said, "Sure, if you are willing to work and to learn something a little different. Take the path up to the orchard, and the others will show you around." I walked the winding trail to the hillside orchard overlooking the rice fields and was astonished by what I saw there. There were trees of all types and sizes, shrubs, vines, vegetables growing in the spaces between the trees, and chickens running everywhere. Hide-san, one of the student workers, greeted me and showed me to the rustic hut that I would share with two others. I spent the next two years in this orchard paradise learning about Mr. Fukuoka's natural farming techniques and the philosophy from which they arose.

By the time I came to Mr. Fukuoka's farm, he had already been practicing natural farming for many years. The story of how he came to be farming that way is both interesting and instructive.

Masanobu Fukuoka grew up in a small village on the island of Shikoku, where his ancestral family had lived for hundreds of years. He worked in the rice fields and in the citrus orchard of his family's farm while he was growing up. Mr. Fukuoka went to the Gifu Agricultural College, near Nagoya, where he studied plant pathology under the tuteledge of the eminent Makato Hiura, eventually taking a job with the Agricultural Customs Office in Yokohama. His primary responsibility was inspecting plants that were entering Japan for diseases and insects. When he was not inspecting plants he spent his time doing research and, as he recalled later, "was amazed at the world of nature as revealed through the eyepiece of a microscope."

After three years there, he developed a serious case of pneumonia and nearly died. Even after he recovered, he spent long hours wandering in the hills contemplating the meaning of life and death. After one of these solitary, all-night walks, he collapsed near a tree at the top of a bluff overlooking the harbor. He awoke to the cry of a heron, and had a revelation that changed his life forever. As he put it, "In an instant all my doubts and the gloomy mist of my confusion vanished. Everything I had held in firm conviction, upon which I had ordinarily relied was swept away with the wind . . . I felt that this was truly heaven on earth and something one might call 'true nature' stood revealed."

He saw that nature is in balance and perfectly abundant just as it is. People, with their limited understanding, try to improve on nature thinking the result will be better for human beings, but adverse side effects inevitably appear. Then people take measures to counteract these side effects, and larger side effects appear. By now, almost everything humanity is doing is mitigating problems caused by previous misguided actions.

Mr. Fukuoka tried to explain his ideas to his co-workers and even to people he met on the street, but he was dismissed as an eccentric. This was in the

1930s, when science and technology seemed poised to create a new world of abundance and leisure. And so he decided to leave his job and return to his family farm to apply his understanding to agriculture. His goal was to create a tangible example of his way of thinking and, in so doing, demonstrate its potential value to the world.

The farm consisted of about one acre of rice paddies where the rice was grown in a flooded field, and a ten-acre citrus orchard. The farmhouse was in the village with a courtyard and a small organic vegetable garden outside the kitchen door. Mr. Fukuoka moved into a small hut in the orchard and spent the next several years observing the condition of the soil and noting the interaction of the plants and animals that lived there. Recalling that time, Mr. Fukuoka said, "I simply emptied my mind and tried to absorb what I could from nature."

Mr. Fukuoka wanted to create a productive environment where nature would have free rein. But where to begin? No one he knew had ever tried that sort of thing before, so he had no mentor to show him the way. He noticed that the plants present in the orchard were limited to citrus trees and a few shrubs, and while some scraggly weeds grew up here and there, the exposed soil had eroded down to the hard, red subsoil. In such a situation, if he simply did nothing, nature would continue in a downward spiral. Because people had created this unnatural condition, he felt a responsibility to repair the damage.

To loosen the soil, he scattered seeds of deep-rooted vegetables such as daikon radish, burdock, dandelion, and comfrey. To clean and enrich the soil, he added plants that have substantial, fibrous root systems, including mustard, radish, buckwheat, alfalfa, yarrow, and horseradish. He also knew he needed green manure plants that fixed nitrogen, but which ones? He tried thirty different species before concluding that white clover and vetch were ideal for his conditions. The roots of the white clover form a mat in the top few inches of the soil so they are effective at suppressing weeds. The vetch grows well in the winter, when the white clover does not grow as readily.

It is important to note that when Mr. Fukuoka carried out experiments such as these, it was always with the goal of solving specific practical problems. They were never done just for their own sake or to try to understand nature—simply to get feedback.

To improve the deeper layers of the soil, he first tried burying organic material such as partially decayed tree trunks and branches that he collected from the surrounding woodlands. Eventually he concluded that this approach gave far too little return for the effort it required. Besides, his goal was to create a self-sustaining system, which, once established, would take care of itself. He decided to let plants do the work instead.

He planted nitrogen-fixing acacia trees here and there among the citrus as well as other trees and shrubs that were hardy and improved the soil deep down. The acacia trees grew quickly, so after eight or nine years he would cut them down and use the wood for firewood and as a building material, leaving the roots to decay over time. As he removed the trees, he planted others in different places so there was always soil-building going on.

Eventually, the soil became deep and rich and the structure of orchard came to resemble that of a natural woodland with tall overstory trees, midsized fruit-bearing trees, shrubs, vines, and a ground layer of weeds, perennials, medicinal plants, mustard, buckwheat, and vegetables. White clover grew everywhere as a permanent, soil-enriching ground cover. By the time I came to the farm, there were more than thirty different kinds of fruit- and nut-bearing trees in the orchard, as well as berries of all kinds, vegetables, and native plants in each of the different layers of the "food forest." There were also chickens and geese running around, a few goats, some rabbits, and bee hives. Birds, insects, and other wildlife were everywhere and shiitake mushrooms were growing on decaying logs which were lined up in shady areas under the trees.

One principle that Mr. Fukuoka followed as he worked out the details of his farming technique was to consider how one could do as little as possible. This was not because he was lazy, but because of his belief that if nature were given the opportunity it would do everything on its own. As he wrote in *The One-Straw Revolution*, "The usual way to go about developing a method is to ask 'How about doing this?' or 'How about doing that?' bringing in a variety of techniques one upon the other. This is modern agriculture and it only results in making the farmer busier.

"My way was just the opposite," he continued. "I was aiming at a pleasant, natural way of farming* which results in making the work easier instead of harder. How

---

* Farming as simply as possible within and in cooperation with the natural order, rather than the modern approach of applying increasingly complex techniques to remake nature entirely for the benefit of human beings.

about *not* doing this? How about *not* doing that?—that was my way of thinking. I ultimately reached the conclusion that there was no need to plow, no need to apply fertilizer, no need to make compost, no need to use insecticide. When you get right down to it, there are few agricultural practices that are really necessary."

When Mr. Fukuoka first inherited the orchard, however, most of the natural systems had been damaged so badly that he had to do many tasks himself that later became unnecessary. Once the permanent soil-building combination of plants had become established, for example, he no longer needed to fertilize. In the early years, until he established a diversity of plants and habitats for insects, he had to grow chrysanthemum plants from which he derived the natural insecticide pyrethrum. He used this to control aphids and caterpillars on his vegetables. Once the soil improved and the natural balance of insects was restored, this too, became unnecessary. Eventually there was very little Mr. Fukuoka needed to do. He scattered seeds and spread straw, cut the ground cover back once each summer and left the cuttings right where they were, replaced some trees and shrubs from time to time, and waited for the harvest.

He got the idea for his rice growing one day when he passed a rice field that had recently been harvested. There he saw new rice seedlings growing up voluntarily among the weeds and straw. Mr. Fukuoka had already stopped plowing his rice fields, but from that time on he stopped flooding the paddies. He stopped growing nursery beds in the spring and then transplanting the young shoots to the main field. Instead, he broadcast the seeds directly onto the surface of the field in the autumn when they would naturally have fallen to the ground. And instead of plowing to get rid of the weeds, he learned to control them by scattering straw and growing a more or less permanent ground cover of white clover. In the end, as with the orchard, Mr. Fukuoka's way of growing rice eliminated all but the simplest of tasks—sowing seeds, spreading straw, and harvesting. He relied on nature to take care of the rest.

When Mr. Fukuoka returned to his family's farm and began practicing natural agriculture, it was with the goal of demonstrating that his way of thinking could be of great value to society. After twenty-five years, the yields in Mr. Fukuoka's unflooded fields equaled or exceeded the top-producing farms in Japan. He also grew a crop of barley over the winter in the same rice fields, and shipped nearly two hundred thousand pounds of mandarin oranges each year, mainly to Tokyo where many people had never tasted naturally grown food before.

Natural farming does not use any of the products of modern technology. While still attaining high yields, it creates no pollution, and the soil improves each year. If Mr. Fukuoka was able to get yields comparable to those of the other farmers in Japan, who use all the latest tools of science and technology, create pollution, grow sickly plants, and ruin the soil, then where was the benefit of human understanding and technology? After just twenty-five years, he had proven his point.

There were no modern conveniences in the orchard. Drinking water was carried from the spring, meals were cooked at a wood-burning fireplace, and light was provided by candles and kerosene lamps. Mr. Fukuoka provided his student workers with thirty-five dollars a month for living expenses. Most of that was used to buy soy sauce and cooking oil, which were impractical to produce on a small scale. For the rest of their needs the students relied on the food that was grown in the fields and in the orchard, the resources of the area, and on their own ingenuity.

Mr. Fukuoka purposely had the students live in this semi-primitive manner because he believed it helped provide the sensitivity necessary to farm by his natural method. He did not pay the students for working there, but no one objected. They felt that living in such an idyllic situation and receiving Mr. Fukuoka's teaching—which was itself freely given—was more than adequate compensation. It has been more than thirty-five years since I lived at the farm. All the work I have done since that time to promote natural farming has been my way of repaying Mr. Fukuoka for what I learned from him.

When I was at the farm, there were five or six of us who stayed continuously for several years. Others would come and stay for a few weeks or a few months and then head back down the mountain. There was a wonderful sense of camaraderie. We would get together in the morning and plan the day's work. Hide-san had been there the longest and had the most comprehensive understanding of the farmwork, so he was informally recognized as the group's leader. The agricultural jobs, such as thinning the fruit crops, cutting back the orchard ground cover, and harvesting might go on for a few weeks or a few months. The daily chores included carrying water, cooking, tending the animals and the beehives, gathering and cutting firewood, making miso (fermented soybean paste) and tofu (bean curd), and preparing the hot bath. Once in a while the huts needed to be repaired or replaced.

Mr. Fukuoka would often work with us instructing us on his techniques as well as practical skills such as making clay seed pellets, growing vegetables in a semi-wild manner, and the proper use and care for tools. He was quite friendly and patient, but his patience ran short very quickly when he saw what he considered sloppy work. Mr. Fukuoka was tireless. Even at sixty-five, he would bound up and down the orchard hillsides like a mountain goat. We all had trouble keeping up with him.

Some days, often on Sunday or during heavy rains, Mr. Fukuoka would gather us together to discuss his philosophy. These sessions were difficult for me. Although I could speak Japanese fluently, I was more fluent in the everyday language we used around the farm. The philosophical and spiritual expressions he used during these discussions were impossible for me to understand. What made this even more frustrating was that Mr. Fukuoka told us over and over that the philosophy was everything, and the farming was merely an example of the philosophy. "If you do not understand the philosophy," he said, "the rest becomes empty activity." So I just did my best each day, and assumed that one day I would get the idea.

One afternoon while we were threshing rice in the courtyard of his home in the village, Mr. Fukuoka emerged from the house with a big smile on his face. He was holding the copy he had just received from the publisher of the Japanese edition of *The One-Straw Revolution*. Mr. Fukuoka had already written several books, but had been forced to self-publish them because he could not find a publisher willing to take a chance on ideas that were so far from the mainstream. Then the first oil crisis occurred in the early 1970s. Japan, as an industrial nation with almost no domestic fuel resources, felt particularly vulnerable. Suddenly everyone was looking for alternatives to petroleum-based production. A publisher finally came to Mr. Fukuoka and asked him to write a book introducing his natural farming method and how he came to be farming that way. He wrote the book in just three months.

After we read the book, the other students and I decided that we would translate it into English and try to get it published in the United States. Mr. Fukuoka's philosophy and techniques were simply too important to languish in Japan where he worked in relative obscurity. I had studied soil science and plant nutrition at the University of California, Berkeley, and knew of the many problems caused by plowing the soil. Many farmers and researchers, even in mainstream agriculture, were trying to develop a no-tillage system for grains and other crops, that would avoid the problems of using so much energy, causing soil erosion and burning out the organic matter in the soil, but no one could figure out how to do it, at least not without drenching the fields with herbicides. So besides the inherent appeal of Mr. Fukuoka's philosophy, I also knew that his twenty-five year example of a high-yielding, chemical-free, no-tillage system would be welcome news in the world of agriculture.

None of us had experience writing, editing, or doing translation, but we would not let that deter us. This was before the advent of personal computers or word processing, so the first order of business was to get the old typewriter that was in one of the huts into usable condition. It had no ribbon, the *d* and the *e* keys were missing, and the carriage had a frustrating habit of sticking on the return. I took the train to Matsuyama city several times to get it repaired, being sure to visit Matsuyama Castle and the public hot springs along the way.

Chris Pearce was a friend I had met during my time living on rural communes. He had grown up in Japan and could speak and read both Japanese and English fluently. He gave us a first-draft translation. But Chris had never been to Mr. Fukuoka's farm and did not have farming experience, so some sections of the manuscript seemed ambiguous or were difficult to understand.

One of the other students living on the farm at the time, Kurosawa-san, had just returned from a yearlong trip to the United States where he toured organic farms. Three or four times a week, after working all day, the two of us sat down with Mr. Fukuoka to clarify these passages. Finally, when we had what we considered a reasonable draft, I was entrusted to go to the United States to find a publisher. That was in 1976.

I managed to get the manuscript to Wendell Berry, who lives and farms in Kentucky. Despite the rough, unprofessional condition of our draft, Mr. Berry liked the content of the manuscript well enough that he took the book under his wing and made sure everything went smoothly for it. He suggested that we use Rodale Press, partly because he did not want the book and Mr.

Fukuoka's philosophy to become characterized only as a "new age" work. He wanted to be sure that the book would end up in the hands of real farmers, because he thought the message would be of benefit to them and could possibly help to reverse the degenerative momentum of modern-day agriculture. Rodale Press published *Organic Gardening* magazine, which at the time had a circulation of more than one million, and had a book club that reached tens of thousands of farmers in the heartland of America.

For the next year or so, Mr. Berry and I worked to smooth out the manuscript and clarify the passages that might be difficult for Americans to understand. The book was published in 1978 and became an instant success. Since then, *The One-Straw Revolution* has been translated into more than twenty-five languages (no one knows exactly how many), all from our English-language edition.

The publication of *The One-Straw Revolution* formed a watershed in Mr. Fukuoka's life. For thirty years before that he had labored in his small village, relatively unknown. After its publication, he became well known and respected all over the world, and he started getting invitations to visit from supporters everywhere.

The first came in 1979 from Herman and Cornelia Aihara, who sponsored a macrobiotic summer camp at French Meadows in the Sierra Nevada. It was the first time either Mr. Fukuoka or his wife, Ayako, had been outside of Japan and the first time either of them had been on an airplane. Over the next six weeks he traveled throughout California, New York, and Massachusetts.

Seven years later he returned to the United States for another six-week trip, which also included visits to Oregon and Washington. Over the last thirty years of his life he also visited India (five times), Thailand (several times), the Philippines, Africa (including Somalia, Ethiopia, and Tanzania), Europe (twice), including a memorable trip to Greece, and China.

When he first saw the condition of the landscape in California he was shocked by how barren it was. Some of that, he noted, was caused by the climate, which lacks the dependable summer rains of Japan, but much of it was caused by careless agricultural practices, poor water management, overgrazing, and overlogging. Eventually he came to refer to this as "California's ecological disaster." After visiting India and Africa, he got an idea of the magnitude of the worldwide ecological crises. From that time on he devoted all his energy to solving the problem of desertification using natural farming.

Mr. Fukuoka believed that most of the world's deserts were created by human activity. These ill-advised actions were based on incomplete human understanding. He felt that the deserts could be revegetated by broad-scale seeding of as many species of plants and microorganisms as could possibly be gotten together. Since conditions had been altered so drastically, it would not do to try to put things back the way they once were. By making the seeds of all the different species of plants and microorganisms available, nature would be able to take the most appropriate course given the present conditions. He refers to this as the Second Genesis. Most important, people's preconceived ideas would be left out of the decision-making process. He believed that plant quarantines should be done away with and large seed banks should be created to facilitate this effort.

Mr. Fukuoka's plan for halting desertification and his thoughts about such things as economics, politics, diet, formal education, the arts, health care, and science, which are all discussed in this book, proceed directly from his core philosophy, which came to him unexpectedly that morning in Yokohama when he was twenty-five years old. He saw nature as a single, interconnected reality with no intrinsic characteristics. He saw time as an uninterrupted moment of the present with past and future embedded within it.

In a futile effort to understand nature and establish a frame of reference, people overlay the reality of nature with notions like north and south, up and down, good and bad, distinguish the various creatures of the world as distinct entities, and create a "human time" based not on what they experience but on clocks and calendars. In doing so, people create and live in a world of human ideas, thereby separating themselves from nature. In the original, absolute world, according to Mr. Fukuoka, these human concepts and judgments do not exist.

For many years the basis of Mr. Fukuoka's worldview was difficult for me to understand, but one day it all became clear. I was walking in a redwood forest on the north coast of California. As I sat down to rest by a small creek, I looked up and saw something I had never seen before . . . the peace and beauty of nature itself. For some reason I was no longer separated from nature by the filter of my own thoughts. Instead of looking *at* nature, I was now within it. Nature had not changed, but my perception was different. I had to laugh. All this time I had been struggling with Mr. Fukuoka's philosophy when what he was getting at was actually so simple and right in front of me all that time.

When I traveled with Mr. Fukuoka on his trips to the United States, people would often ask him if natural farming could be mixed with conventional or organic practices. He was adamant that you could not do that. Now, finally, I understood why. One either lives in the absolute world of nature, or in the fantasy world of human thoughts. There is no middle ground.

Given Mr. Fukuoka's worldview, it is not surprising that his natural farming practices, and the plan he advocates for revegetating the deserts of the world, seem to go against conventional wisdom and current scientific thinking. It is a truly visionary, outside-the-box strategy that he believed would return humanity to its correct relationship with nature and heal the confusion and suffering of the human heart. His goal is no less than re-creating the Garden of Eden where people would live together in abundance, in freedom, and in peace.

Larry Korn
Ashland, Oregon
2012

Editor's Notes

THE JAPANESE EDITION of *Sowing Seeds in the Desert* was first published in the mid-1990s. A second, slightly revised version came out a few years later. Mr. Fukuoka had a particular fondness for this book because it explains the heart of his philosophy and discusses what became his passion over the final thirty years of his life—applying natural farming to stopping the spread of the deserts, and to revegetating the earth. We are pleased to make this important work available to English-language readers.

A literal translation from one language to another presents challenges arising from both linguistic and cultural differences. Translating from Japanese to English presents characteristic difficulties of both kinds.

One major linguistic concern is that Japanese uses many passive forms in sentence construction, while English has a preponderance of active forms. A literal translation from the Japanese will tend to create a tone of formality and indirect expression in English which does not exist for Japanese readers. In adapting the original translation, I have re-worded numerous passages from passive to active voice so that they preserve the friendly, direct personality of the author.

The most conspicuous of the cultural difficulties is that the Japanese way of telling a story or developing a complex argument is different from the approach that is generally taken in English. In Japanese, the author typically begins with the theme or the point he wishes to make. Then he offers an anecdote or an argument that helps to tell that story or bolster the point before returning to the theme, which is restated. Then the author goes on another loop, again returning to the theme. One might say that these side stories or arguments form the petals of a flower with the theme as its center. This helps establish the sincerity and conviction of the author in the Japanese reader's mind. However, this circular style can quickly become tedious to English-language readers who are used to a more linear approach. Therefore, some sections of the book have been restructured so that the order of presentation of Mr. Fukuoka's ideas is more natural for an English-speaking audience.

In both cases—the active *versus* passive sentence construction, and the circular *versus* direct approach to telling a story or making a point—the intent has been to make the book read as smoothly and as clearly for English-language readers as the original is for Japanese readers.

In some places, Mr. Fukuoka uses terms or expressions that require a cultural context English-language readers may not be familiar with. In these cases, additional explanation has been provided in notes. Sections of other works by Mr. Fukuoka, as well as parts of conversations with him, have been included in the text.

The publisher and editor wish to thank Michiyo Shibuya and Wayne Olson for their assistance in the preparation of this book.

L. K.

About the Illustrations

ALL OF THE ILLUSTRATIONS in this book were created by Masanobu Fukuoka using a brush-and-ink technique called *sumi-e*. Some were done as autographs in copies of *The One-Straw Revolution*. Others were created on the spur of the moment and simply given away. He carried his art supplies everywhere he went during his travels.

Each drawing is intended to explain his philosophy in one way or another. They also demonstrate the playful, fun-loving side of Mr. Fukuoka's personality.

FRONTISPIECE
This drawing shows the "cave of the intellect," which is described on page 12.
Without doing, without thought, without knowledge.

TITLE PAGE
All one sky.

DEDICATION PAGE
The Buddha sits holding a child and a string tied to a balloon. The balloon represents the human world of relative thinking. The Buddha and child, who are not in the balloon, represent the absolute world of nature.
Everything is inside the balloon.

Introduction

Two people are sitting by the fire inside a clay pot. The pot represents the world created by human thoughts. The three characters around the people are *wind, light, fire*. The character in the smoke that has managed to rise out of the pot is *mu*, or emptiness. The third person, who is not inside the pot, is relaxing and enjoying himself.

The hearth is the universe /
the universe is also a mid-day dream.

Recurring Image to Begin Each Chapter's Text

This image, which is repeated at the beginning of each chapter, shows a cooking pot hanging over a campfire.

Chapter 1

On the right, people are enjoying themselves in nature. On the left they toil in the "cave of the intellect." The contrast is intentional and striking.

The hearth is the universe.
People do not cultivate the soil /
they cultivate knowledge of the cave.

Chapter 2

There is no wide or narrow on the earth /
there is no fast or slow in the sky.

Chapter 3

This drawing was inspired by Mr. Fukuoka's experience at the summer camp at French Meadows in the Sierra Nevada Mountains. It shows people camping out, delighting in the forest, the river, and the fresh mountain air. The cooking pot over the fire pit is suspended by the moon.

In the place where there is nothing, everything exists.

Chapter 4
Here, people are carrying out various forms of human activity with the idea that they are accomplishing something. The last line refers to the figure on the left who is on his own mission to repair the damage of that activity.

Originally there was no east/west, high/low,
slow/fast, easy/hard, love/hate.
This person is sowing seeds in the desert.

Chapter 5
Say, mister cricket, your house is inside a cucumber/ my house is in the blue sky. My light comes from the moon and the stars, so I don't really need a chandelier.

Chapter 6
This illustration is described on page 143.

Appendix Opener
Two people are sitting beside an oversized cooking pot. The literal translation of the first line implies a sort of drudgery, but the Japanese meaning is much brighter. A simple life with its everyday activities has a joy that is its own reward.

Eat, sleep, eat, sleep / nothing to do for 100 years.

人は大地を耕さず
洞窟の闇を耕す

炊爐
邊
宇宙

CHAPTER ONE

The Call to Natural Farming

FIFTY YEARS AGO, I had an experience that changed my life forever. I was twenty-five years old at the time. After graduating from Gifu Agricultural College in plant pathology, I took a job with the Plant Inspection Division of the government's Customs Bureau inspecting plants that were coming into or leaving the country. A believer in science, I spent most of my time peering into a microscope in the laboratory, which was located beside a small park in the Yamate area of Yokohama.

After about three years there, without warning, I was stricken with acute pneumonia, which brought me face-to-face with the fear of death. After I recovered, I began questioning the meaning of human existence. Immersed in my distress, I wandered through the hills, day and night.

After one full night of aimlessly wandering, I collapsed, exhausted, at the foot of a tree on a bluff overlooking the harbor. I sat in a daze, drifting in and out of sleep as dawn approached. Suddenly the piercing cry of a night heron awakened me as if from a dream. All the confusion, all the agony that had obsessed me disappeared with the morning mist. Something I call "true nature" was revealed. I had been transformed, body and soul. The first words that rose to my lips were, "There is really nothing at all." I looked around in joyful amazement.

The peaceful beauty of the world became vividly apparent to me. I was overcome with emotion and reduced to trembling. I had been foolishly searching for something when it had been there, right in front of me, all this time.

The sparkle of the morning dew, the green of the trees bathed in morning sunlight, the delightful chatter of birds gathered in the dawn . . . what a wonder it was that I, too, was able to take my place in this realm of freedom, this world of ecstasy.

I saw nature directly. It was pure and radiant, what I imagined heaven to be.

I saw the mountains and rivers, the grasses and trees, the flowers, the small birds and the butterflies as if for the first time. I felt the throbbing of life, delighted in hearing the songbirds and the sound of rustling leaves. I became as light as the wings of a dragonfly, and felt as if I were flying as high as the mountain peaks.

The question is why, on that occasion alone, did the world that I was used to seeing every day appear so fresh and new and move me so deeply? To be honest, my mind at the time was not in its usual state. I had reached the point of mental and spiritual exhaustion from my sickness and had no strength of will left. In the soft, tranquil air of daybreak, I was neither waiting for the dawn nor looking about for anything in particular. All of a sudden, with that heron's single cry, I was awakened. My heart opened and I was unable to stop my ceaseless flow of tears.

In a single leaf, a single flower, I was moved to appreciate all the beautiful forms of this world. What I saw was simply the green of the trees sparkling in sunlight. I saw no deity other than the trees themselves, nor did I perceive a spirit or soul of vegetation hidden within the trees. When I viewed the world with an empty mind,* I was able to perceive that the world before me was the true form of nature, and the only deity I would ever worship.

The experience I had that morning is indelibly impressed in my mind, its freshness undimmed even today. But realistically, I could not expect those feelings to last forever.

I was brimming with self-confidence, convinced that I had come to possess the wisdom of truth. I felt that I could solve all the problems of the world. But as the days passed, the whirlwind of emotion abated, and the true form of nature, which had been so clear to me, gradually receded. While my viewpoint might have changed, it did not necessarily mean that I myself had undergone a fundamental change.

It took a few years to be able to integrate this new understanding into my daily life, but the first thing I did was to quit my job with the Customs Bureau. I set out from the Boso Peninsula in Chiba Prefecture and

---

* This expression refers to an egoless state in which there is no separation between the individual and the totality of existence.

wandered westward for a month or two, stopping here and there along the way. I danced in the beauty of nature as I traveled, reaching Kyushu as winter approached.

During my travels I talked with a number of people about my realization. In those discussions I could see that my ideas were at odds with the way the rest of society was thinking at the time. When I said that humankind lives in an unreal world separate from nature, I was told that I was simply deluding myself. Eventually, since I could find no words to adequately express what I had seen that morning, I learned that it was better to remain silent. As time went on, that pure vision of nature I had experienced grew weaker and weaker.

Had I been the sort of person to immerse myself in religious discipline, I suppose I would have renewed my vows. But being at heart a carefree individual, I chose to pursue a life of farming, separating myself from the commotion of modern society.

My Return to Farming

After spending some time living in a lakeshore hut outside the town of Beppu, I returned to my parents' farm[*] in Ehime Prefecture on Shikoku Island. It was the

---

[*] Mr. Fukuoka grew up in a small farming village sixteen miles from Matsuyama city. The Fukuoka family had been settled there for hundreds of years. The farm consists of about one acre of rice and barley fields, and a ten-acre citrus orchard situated on a hill overlooking Matsuyama Bay. It takes about twenty minutes to walk from the fields to the orchard. As the eldest son, Mr. Fukuoka eventually inherited the farm and, with his wife, Ayako, raised five children there.

spring of 1938. I began living alone in a hut in the citrus orchard, and decided to start a natural farm to express what I had seen that morning in physical form. Most importantly, I wanted to demonstrate how my ideas could be of practical benefit to society.

At first, I was not sure how to go about it, but I was resolved to making the orchard at the top of the hill into a paradise. My idea was to let nature have a free hand. At the time, the citrus trees my father planted had been carefully shaped, but because I failed to prune them—with the idea that they would revert to their natural form—the branches grew out every which way, insects and diseases appeared, and before I knew it I had wiped out more than two hundred trees.

This first experiment, simply doing nothing, was a magnificent failure. It was not natural farming; it was abandonment. But I was pleased that at least I had learned from that disaster the difference between nonintervention and taking human responsibility.

My father was worried about me, but I persisted. Unfortunately, however, the tide of human history carried me in an unexpected and unwanted direction. The skies over Asia had become dark and dangerous. Soldiers were marching off to war, and the drumbeat grew louder and louder. My surroundings lost their serenity. It was no longer possible to turn my back on the world and live alone and carefree in my hillside hut.

My father, who was mayor of the village at the time, urged me to find work somewhere, anywhere, even at an agricultural experiment station. It was a time when the government and the farmers were pouring all their efforts into increasing food production, so I decided, after an absence of five years, to return to work for the government as a plant pathologist. I obeyed my father's wishes and left the farm.

Here is something I scribbled in my hut at that time.

> *Wishing to cultivate the earth,*
> *I cultivate understanding.*
> *In vain I wield my hoe*
> *And sharpen my sickle.*
> *The earth languishes, grasses and trees wither,*
> *Gazing at heaven and earth and heaving a long sigh,*
> *I am filled with despair.*
> *When will*
> *That Garden of Eden*
> *Bloom again?*

That small hermitage, Musoan,* commemorates the birthplace of natural farming, but there is no trace of it left today. That simple mud-walled hut has, over time, simply returned to the earth.

Challenges During Wartime

And so I spent the years of World War II at the agricultural experiment station in Kochi Prefecture on Shikoku. Looking back, I was completely irresponsible while working there. Instead of concentrating on my research about blight and insect damage in plants, I argued with my co-workers, asking them to consider

---

* Translates to "the Hermitage of No Thought."

my ideas denying the validity of science. Or else, under the pretense of investigating plant diseases and insect pests, I rambled through the mountainous interior of Kochi doing my own research about nature. I had one foot in natural farming and the other in the world of science. My life was full of contradictions.

As Japan's policies were leading the country into war, I deeply regretted that I could do nothing to stop it. One Sunday, five or six young soldiers from the nearby air force unit came to visit on their day off. I welcomed them, of course. I wanted to offer them something, but there was no food. So they just spent the entire day relaxing in my upstairs room and then went back to the base. The following morning they disappeared into the southern sky. It still breaks my heart to recall the boyish faces of those young men.

As Japan entered the last year of the war, I, too, was drafted. Luckily the war ended soon after and I was spared. I received my discharge and gratefully returned home. I was finally back working in the fields again, savoring the joy of simply being alive. I especially remember the buoyant sound of the threshing machine as I treaded it for the first time after a long absence.

Since that time, I have been a farmer, never veering from the path of natural farming. I experienced many failures as I developed my natural method for growing rice and winter grains in an unplowed field, broadcasting the seeds right onto the surface of the soil, but I persisted. I was also interested in creating a natural orchard without cultivating the land, using chemicals, or doing hand weeding. Before I knew it, forty years had passed. This is not to say that I was a particularly industrious farmer. Actually, I was aiming at a do-nothing* method of farming.

Yes, many years have passed since the inspiration of my youth. I am a white-haired old man now. When I look back on the postwar period, it seems so long ago and yet it seems so short. As I consider the few years that remain to me, I wonder if I have done all that I could have.

The road I have taken has been a single path. The starting point of this journey, it turns out, is also its end.

The True Meaning of Nature

I spent many years of my youth foolishly searching for something I "should" have been doing. Instead, I should have entrusted everything to the flowers blooming in the meadow. Even if people do nothing at all, the grasses and trees and the songbirds will live on.

The poet Basho composed the haiku, "Ah, how sacred / the light of the sun / on young green leaves." Indeed—I can clasp my hands in reverence and kneel before the daikon flower. Even if I cannot make a poem as beautiful as Basho's, my heart is singing, "Oh, the whiteness of the daikon flower / the radiance, the splendor!"

The sad truth is that for much of my youth, I, too, felt estranged from nature. But now I just take a single

---

* With this expression Mr. Fukuoka draws attention to his method's relative ease. To be sure, this way of farming *does* involve hard work, especially at the harvest, but far less than other methods. His aim is to avoid *unnecessary* work, especially work that was created as an adverse side effect of previous actions.

flower in my hand and converse with it. I have finally learned that, although nature does not reach out to people directly, people can always approach nature and seek salvation that way.

In ancient times, I would like to think that people must have made drawing close to nature the most important goal in their lives.

Once long ago, when I was in the mountains, I unconsciously wrote, "The mountains, rivers, grasses and trees are all Buddha," on a piece of wood. At other times I would suggest that "God" refers to the absolute truth that transcends time and space. Perhaps an even better description, I sometimes thought, was Lao-tse's[*] term "The Nameless." I was really just struggling with words. Actually, I think people would be better off without words altogether.

The Errors of Human Thought

Because of the rapid development of modern science, the Asian tendency to live quietly and view the world as transitory is disappearing. The new trend is toward glorifying modern civilization and the idea that the material is almighty.

Within the history of the development of Western science, the epoch-making discoveries that have had the greatest influence on the human race are (1) the theory of biological evolution advanced in Darwin's

---

[*] A philosopher of ancient China, Lao-tse is widely considered the founder of Taoism.

The Origin of Species, (2) Newton's universal gravitation and Galileo's heliocentric theory, and (3) Einstein's relativity theory of the universe.

Darwin started with the progress of human beings on earth and followed the traces of the origin and development of living things, eventually determining that living things have evolved. The idea that human beings must continue to develop became firmly rooted in people's minds.

Newton, seeing how an apple fell, discovered the law of universal gravitation and laid the foundation for modern physics. Galileo understood that the earth was round and, when placed on trial by the church, did not falter in expounding his theory that the earth revolves around the sun. By denying the fallacy that the heavens revolve around the earth, he dealt a serious blow to the theory of divine creation.

By establishing the relativity theory of the universe, Einstein propelled the human race into the space age. To everyone's amazement, he concluded that there is no speed faster than the speed of light, overturned the commonly accepted belief that light always travels the shortest distance in a straight line, and proposed the new theory that light is curved.

In addition, Einstein said that light waves, radio waves, and electromagnetic waves are all the same and that they travel through space at a fixed speed, regardless of their length, without accelerating. From his formula that mass and energy are equivalent, the launch of man-made satellites and space vehicles became possible.

The religion of Buddhism, however, rejects knowledge acquired through the human intellect as nothing more than illusion. Some Western myths are also skeptical of human knowledge teaching that since Adam and Eve ate the fruit of the Tree of Knowledge, humankind has been banished from the Garden of Eden.

Western philosophy has been divided on this issue, however. Socrates, for one, started with the assumption that human beings know nothing. Descartes, on the other hand, declared, "I think, therefore I am." Beginning with that conviction, that people can and do know themselves, he made human judgment his standard, established rules for the physical world, and began analyzing its properties.

Scientists have historically assumed that it is acceptable to control nature using human will. Nature is seen as the "outside world" in opposition to humanity, and this idea forms the basis of modern scientific civilization. But this fictitious "I" of Descartes can never fully comprehend the true state of reality.

Just as human beings do not know themselves, they cannot know the other. Human beings may be the children of "Mother Nature," but they are no longer able to see the true form of their mother. Looking for the whole, they only see the parts. Seeing their mother's breast, they mistake it for the mother herself. If someone does not know his mother, he is a child who does not know whose child he is. He is like a monkey, raised in a zoo by humans, who is convinced that the zookeeper is his mother.

Similarly, the discriminating and analytical knowledge of scientists may be useful for taking nature apart and looking at its parts, but it is of no use for grasping the reality of pure nature. One day scientists will realize how limiting and misguided it is to hack nature to pieces like that.

I sometimes make a brush-and-ink drawing to illustrate this point. I call it "the cave of the intellect." It shows two men toiling in a pit or a cave swinging their pickaxes to loosen the hard earth. The picks represent the human intellect. The more these workers swing their tools, the deeper the pit gets and the more difficult it is for them to escape. Outside the cave, I draw a person who is relaxing in the sunlight. While still working to provide everyday necessities through natural farming, that person is free from the drudgery of trying to understand nature, and is simply enjoying life.*

Ironically, nature is also being damaged by people who pride themselves on following a path of moderation, who may think they have nature's best interest at heart.

These well-meaning people, known for their compassion and practical sense, might say:

"Human beings have lived in nature for thousands and thousands of years, sometimes joyfully, sometimes filled with sorrow. Isn't that the essence of the relationship between people and nature? Isn't it simplistic to see nature as only filled with truth, good, and beauty while seeing human beings as insensitive and ignorant?" At first glance, this opinion seems sensible, and appears to be an objective point of view. But these goodhearted people have not escaped the realm of relative thinking.

Seen from a nonrelative perspective, nature transcends beauty and ugliness, good and evil. Whether

---

* This drawing is reproduced as the frontispiece for the book and is also part of the illustration before Chapter One.

we see this world as filled with contradictions, or as existing in perfect harmony, is determined by whether we analyze it using our intellect, or grasp the entirety of nature without making any distinctions at all. It is only by doing the latter that we can see nature's true form.

No God or Buddha Will Rescue the Human Race

The destruction of nature will lead to the destruction of the human race, but many people seem to be convinced that even if humans should disappear, they will be brought to life again by the hand of their god. This idea, however, is nothing more than fantasy. The human race will not be born again. When the people on the earth have died out, there will be no God or Buddha to rescue them.

People do sometimes sense the sacredness of nature, such as when they look closely at a flower, climb high peaks, or journey deep into the mountain. Such aesthetic sense, love, receptivity, and understanding are people's most basic instincts—their true nature. These days, however, humans are flying in a completely different direction to some unknown destination, and they seem to be doing it as rapidly as possible.

Perhaps the people who most easily perceive that nature is sacred are a few religious people, artists of great sensitivity, and children. With their compassion they often perceive, at the very least, that nature is something beyond human invention and that it should be revered. The poets who write about nature, the painters who turn it into works of art, the people who compose music, the sculptors ... I would like to believe that they are the ones drawn to what is truly meaningful.

But if an artist's understanding of nature is unclear, no matter how keen his sensitivity, no matter how excellent his power of expression, no matter how refined his technique, he will eventually find himself lost.

The Dragonfly Will Be the Messiah

There has never been a generation like the present where people's hearts are so badly wounded. This is true of every area of society—politics, economics, education, and culture. It is reflected in the degradation of the environment, which comes about through the material path humanity has chosen. Now we have the ugly sight of industry, government, and the military joining forces in the struggle for ultimate power.

In the present age of disintegration the various religions of the world, old and new, large and small, are becoming very active. Indeed, whenever the world has fallen into disorder, religious movements have flourished.

Let me give one example of a religion that promises wealth and good fortune. A young man who was worshiped as the founder of a new religion in Kobe came to my farm with ten or so of his disciples. This fellow told me that he had received special training to transform him from an ordinary religious person to the founder of this new religion. He learned such things as physiognomy,* mind reading, fortune-telling, palmistry,

---

* The art of judging human character from facial characteristics.

divination, hypnotism for healing disease, exorcism, and various ways of communicating divine messages, such as writing in sand. He told me in great detail about the schemes he had used to get believers into the palm of his hand, starting with tricks for determining a potential believer's character flaws and problems. This, he said, would help him attract new followers.

This is only one type of many religious imposters who hold both the deities and the people captive and run around acquiring believers in order to make money and gain power. But many of them are popular and well regarded, and would not seem to be the stereotypical image of an imposter. This paradox leads me to reflect on how human beings are nothing more than animals dancing to a tune piped by their own ideas.

I look forward to the day when there is no need for sacred scriptures or sutras. The dragonfly will be the messiah.

A Life of Natural Culture

When I mention that human society is on the wrong path, I often hear the retort, "Then show me a better one." Because it does not have a name yet, I will refer to it as "natural culture and community." Natural culture is simply a way of life in which people enjoy the truth and beauty of nature, a life in which people, with freedom in their hearts, climb mountains, play in meadows, bathe in the warm rays of sunlight, breathe pure air, drink crystalline water, and experience the true joy of life.

The society I am describing is one in which people will create a free and generous community.

Once the primal source of nature is destroyed, however, it will no longer be able to restore itself, and this image of a natural culture will become obsolete. Indeed, many species of plants and animals become extinct each day, and the meaning of the disappearance of one bird or one plant is not just the death of that bird or that plant. It is of grave significance to us all. It is connected with the destruction of the harmony of all living things.

If humanity can regain its original kinship with nature, we should be able to live in peace and abundance. Seen through the eyes of modern civilization, however, this life of natural culture must appear to be monotonous and primitive, but not to me.

There are many other people besides me who question the path of modern society. They are filled with foreboding, wondering whether or not we can solve, or somehow evade, the current environmental crisis. There are even many scientists who believe that the long-term sustainability of life on earth, from the standpoint of the natural environment and its resources, will be decided in the next twenty or thirty years. It is these people to whom I speak directly.

We must realize that both in the past and today, there is only one "sustainable" course available to us. We must find our way back to true nature. We must set ourselves to the task of revitalizing the earth. Regreening the earth, sowing seeds in the desert—that is the path society must follow. My travels around the world have convinced me of that.

碧空に遅速無し

大地に
広狭
無く

CHAPTER TWO

Reconsidering Human Knowledge

THE HUMAN RACE first appeared on earth a couple of million years ago, and we began living what people generally consider "civilized" life several thousand years ago. In Japan, however, people began living a "civilized" life only in the last few hundred years. With the rapid development of science and technology, without our even having time to figure out where we are going, it seems that modern civilization has peaked and that disorder is emerging on a worldwide scale. But is the conclusion to this unfortunate chapter in human history inevitable? What will happen to the world in the future?

The Birth of Discriminating Knowledge

From the time that a child sees the moon floating in the sky and says, "I see the moon," human knowing begins.

When a child first becomes aware of the moon, that child is simply filled with wonder. Then after a period of time the child learns to discriminate between a subject, "I," and an object, "the moon." The child comes to know the thing called the moon as "other." So even in the structure of human language, human beings are taught to set themselves apart from nature. The intimate and harmonious relationship between people and nature that once existed—which we can see in children's instinctive wonderment—becomes cold and distant.

Even if we say that we know the green of nature, this is merely the understanding that discriminates green from other colors. If we have not grasped the *intrinsic* greenness of the grasses and trees, which originates with the life at their core, we cannot say that we really understand what true green is. People simply believe they understand by making a distinction based on the outer appearance.

If knowledge of a whole (one) is broken into two and explained, and then these are divided into three and four and analyzed, we are no closer to understanding the whole than we were before. When we do this, however, we come under the illusion that knowledge has increased. But can we say that by endlessly repeating our divisions and analyses and then gathering up all the fragments, we have advanced human knowledge in any meaningful way?

No matter how much we accumulate, synthesize, and make judgments, this effort is not useful for clarifying the true state of things. Furthermore, it throws us into confusion. Once people create a mental image of the "moon," the moon takes concrete form and we want to "know the moon," and then discover more and more about the moon. This desire eventually leads to actually traveling to the moon and retrieving stones to bring back to earth for further research. The next thing you know we are building space stations, and who knows what else might follow.

With increased "knowledge" comes an increased desire for *more* knowledge, and then people work and work to invent machines to help them achieve even greater knowledge. But even if "proof" is found, only more questions will arise from that so-called proof. The desire for knowledge becomes endless and we lose sight of our place in the world. In the end, the true essence of the moon is more clearly seen through the eyes of a child.

Darwin's Theory of Natural Selection

Darwin's theory of natural selection gives a good illustration of the principle that although the discriminating knowledge of science may seem useful for taking nature apart and analyzing the fragments, it is of no use for grasping the reality of nature.

The Darwinian theory of evolution breaks down the 4.6 billion years since the earth came into being, observes the living things that came into existence at certain times and places, and examines their mutual relationships. Based on this, the diversification and systematic development of the organisms are inferred and classified, and this is all developed into his theory of evolution.

In other words, when oxygen and water formed on the planet—which, in the beginning, was apparently an inorganic mass—primitive life-forms began to appear. These evolved and new life-forms appeared. At first, extremely simple microorganisms such as fungi and bacteria were born; then they developed and branched out, organisms of other forms were born, and gradually the more complicated higher plants and animals began to multiply upon the earth.

The idea is that the life-forms on earth came into being in sequence, along with the development of the earth itself. Various life-forms appeared and lived as part of the food web, but only those that successfully adapted to their environment survived. This is known as the theory of natural selection, the theory of the most adaptable, and sometimes popularly referred to as the survival of the fittest. Among all the life-forms, those that are selected by nature and survive the struggle for existence obtain the right to live and reproduce.

One question I have about this theory is: What basis was used to determine which species are higher or lower, and which are strong or weak? To decide that the phenomenon of the survival of the fittest is the providence of nature and that people are the highest, most evolved species seems to reflect more the strongman logic of human beings than the true state of nature. Actually, no one can say which species is the strongest because all living things depend on one another to survive, reproduce, and eventually decompose, so life can go on for all.

It is true that all forms of life—by necessity and by natural design—consume one another to live, but they do not intentionally bring about another's extinction, systematically deprive other species of their source of food, or create factions and wars. The same cannot be said for human beings.

In nature's cyclical rhythms, there are no grounds for the discriminatory view that underlies Darwin's view of superiority and inferiority that deems single-celled organisms as lower, and more complicated life forms as higher. It would be more appropriate to say we are all one continuous life-form.

Whether you see differences between butterflies and moths, dragonflies and fireflies, depends on whether you are looking at the big picture or the small one. In the eyes of children, frogs, fish, birds, and squirrels all appear to be the same friends, but the eyes of adults are drawn to the differences in appearance and form, and so they appear to be different animals.

Viewing the world macroscopically or microscopically simply means that different scales are being used. Depending on whether you use human time, measured in minutes and seconds, or the eternal time of the Ganges River—or whether your field of vision is as narrow as the inside of a box or as wide as the universe—the appearance of the world changes completely.

Similarly, there is a common belief that there is a great difference between living and nonliving things, but even that distinction occurs only because of one's perception.

Understanding True Time and Space

The commonsense understanding of time is that it is a direct, linear flow from the past through the present to the future. Darwin's theory of evolution is based on this idea of human, historical time. The organisms are classified into fragments in time and location, and systematized. This systemization emphasizes the differences among them. Species that were originally brothers and sisters became divided in people's thinking, by being made into distinctly different things.

Transcendent time, or time as it exists in nature, is a continuous moment of the present. When one sees and operates within that time and space, it is the *unity* of all things that is perceived.

The idea of time that people generally accept came into being with the invention of the calendar and the clock. But a clock, with its needle going around a series of numbers, is just a means of counting.

Time does not simply flow mechanically in a straight line in a fixed direction. We could think of time as flowing up and down, right and left, forward and backward. As time develops and expands, multifaceted and three-dimensional, the past is concealed within the instant of the present, and within this instant of time is concealed the eternity of the future.

It is easy to liken the flow of time to the flow of a river. But even the phenomenon of water flowing in a river presents challenges of perception. When you stand on a riverbank and look at the water, you can clearly see that the water is flowing in one direction. But if you are in a boat moving at the same speed as the water, the river does not seem to be flowing at all; rather the riverbank appears to be moving upstream. As a Zen master once said, "The river does not flow. The bridge flows."

The Rising and Sinking of Genes

Many years ago, when I was a young man working as a microbiologist, bacteriophages—viruses that infect and consume bacteria—were discovered. They were said to be the first-known nonliving things that reproduced. I was quite interested in them as some intermediate matter between living and nonliving things. (By now, of course, research on viruses has further blurred the boundary between living and nonliving things.) When looked at from the standpoint of their elemental particles, the distinction between animate and inanimate, living and nonliving things, plants, animals, viruses, gases, and minerals becomes inconsequential.

The fact that the structure of DNA, and the genetic code by which this genetic material is translated into protein, is the same for all living things indicates to me that all living things are fundamentally related. These proteins could be seen as the liaison between living and nonliving forms and so play the role of threads tying the living and nonliving together.

So, while there seems to be a great difference between plants and human beings, their genetic material is the same. Whether something becomes a plant or a human being is just a question of whether or not the genetic factor for greenery surfaces or sinks. In fact, only a small portion of the combinations of the four genetic elements have successfully come into being, while most of the other combinations have been lost or are dormant.

The reason that there are so few intermediary forms between species and that we cannot find fossils of them is not necessarily because they did not exist. Rather than saying that the genes that become intermediate species have not functioned at all, we can only assume that even if they *were* born, they died in infancy and did not come to the attention of human beings. That is why different, seemingly disconnected species and varieties are left.

It is like an acacia tree that produces millions of seeds, each with its own distinctive genetic makeup. Few of them actually germinate and thrive. After ten years, perhaps only one or two trees will survive as the descendants of the parent tree, but all the other trees that did not survive were also possible.

Take the islands of Japan's Seto Inland Sea* for another example. The many islands in this sea have various shapes, they have been given different names, and they appear to be separate islands, but they are all connected at the bottom of the sea, so you could consider them all the same single "island of Japan." On a larger scale, of course, one could say that they are connected to all the islands and continents of the world by being part of the earth's crust. In the same way, the animals and plants living on earth appear to be different, but they also are all connected at the base. Whether the genes survive, surfacing like islands floating in the Inland Sea, or do not, sinking beneath the water, depends on the arrangement of the genes and on the constant rearrangement of subatomic particles.

---

* The small sea among the main Japanese islands of Honshu, Kyushu, and Shikoku.

An Alternative View of Evolution

Perhaps it went like this: The Creator rolled the seeds of every living thing into clay pellets and gave them to messengers to scatter randomly. Some seeds were programmed to become active soon after the birth of the earth. Others were programmed to thrive in water. Some were suited to the mountains, some to the deserts. The seeds that were designed to become human beings were made to come to life during the later ages of the earth.

Millions of seeds were broadcast at one time, and the living things of the world took on various forms. Some became microorganisms, some became green plants, and some became animals that could run around.

Scientifically, we could say that these seeds germinated when conditions were right for their germination, and only those that took forms suited to their living environment developed and survived. Those seeds that fell into the sea became seaweed, coral, and sea anemones, others became shellfish and shrimp, and all lived together.

The seeds scattered in marshes became cattails, some changed into catfish or eels, and in some cases the same seeds became frogs, turtles, or snakes. Living things bearing similar genes but manifesting differently became forest trees, while others became the birds that lived in the forest trees. At the same time that the vegetation on earth increased and large trees grew in abundance, large animals such as tigers and elephants made their appearance. Microorganisms, plants, and animals are all genetic siblings, of course, but each appeared in a different costume.

In some ways, this may seem like a distorted dramatization of Darwin's theory of evolution. The big difference is how we think about the passage of time. In my mind, millions and millions of years appear as a momentary flash. Consequently, the innumerable varieties of living things have not originated at different times in different places, but rather at the same time, and in the same place.

Nature is one body. We can say that while human beings and insects are part of nature, they also represent nature as a whole. And if that is so, when we harm plants, microorganisms, and insects through large-scale conventional agriculture (to use just one example), we are harming humanity as well.

I would like to propose a dharma wheel theory of biological development as an alternative to Darwin's flat, single-plane theory of natural selection. I will call it the Dharma Wheel Theory of Flux in All Things. The dharma wheel can be seen as a representation of natural law. Nature expands in all directions, three-dimensionally, and at the same time, as it develops, it converges and contracts. We can see these changes of expansion and contraction as a kind of wheel. It is like the universe—three-dimensional, always expanding and contracting, spinning in space, and heading in an unknown direction.

At the creation, along with the birth of the rest of the universe, the earth and all the living things on it were born as a single, unified body with a common fate. Everything regarding the roles, the aims, and the work of each of them originated and was concluded in the same instant. All things were designed so that one is many, the individual is the whole, the whole is perfect, there is no waste, nothing is useless, and all things perform their best service.

There is another aspect to this dynamic, spinning, expanding, and contracting three-dimensional and multifaceted dharma wheel. Its center, the hub, is forever motionless and forever one. Instead of seeing the distinctions among the things of this world, if we look at the base, it is all one, and the purpose of all things is the same.

Naturally Occurring Hybrids in My Rice Fields

Ten or so years ago, I tried crossbreeding nonglutinous rice from Burma with Japanese glutinous rice in order to develop a new variety suited to natural farming. The two rices are extremely different in character, but the genetic characteristics of the parents were intermingled, so I wound up getting twenty, thirty, sometimes several hundred different varieties. When I arranged these varieties, I noticed that their characteristics formed a continuum.

There was nonglutinous rice close to glutinous, intermediate forms, glutinous rice close to nonglutinous, and some in which glutinous and nonglutinous grains were mixed together on a single head. Some rice plants reached a height of only ten or twelve inches, while others were giants more than five feet tall. There was also a succession of different colors of rice—white, red, and blackish brown. Some were flavorful, some not, some powdery, some sticky. In this situation, it was impossible to say which was good and which was bad. I also came to question the value of distinguishing between glutinous and nonglutinous rices, and even between paddy and dry field culture.

After years of crossbreeding rice in my fields, however, I finally concluded that on a natural farm, people do not need to create new varieties by artificial crossbreeding at all, since the insects that most people consider as harmful were creating new varieties on their own.

In my rice fields, I noticed that after locusts and other insects had chewed round holes in the rice grains just as the heads were sprouting, slugs, snails, cutworms, and other creatures came along and crawled over the grains at night. They ate down to the stamens in the holes, after which windblown pollen from other varieties adhered and achieved fertilization. In other words, rice, which is said to be self-pollinating, can also be pollinated by other plants, and in this way new varieties arise naturally.

In a conventionally farmed paddy field sprayed with insecticide, natural hybrids do not occur. On a natural farm, however, they can easily survive, and there are many chances for new varieties to appear. In the end, there is no need for people to imitate nature by carrying out artificial crossbreeding. It is all being done for them.

Alongside crossbreeding rice with rice, I experimented with crossbreeding rice with weeds such as deccan grass (*Echinochloa colona*) and foxtail (wild rye) and was thinking that if that went well, I would try more combinations with foxtail millet and Chinese millet, but my original purpose was not to study rice for its own sake. I was really just amusing myself by going in the opposite direction of what was being recommended by agronomists at the time. I was doing a reverse breeding in search of atavisms: potentially valuable species that had been lost over the centuries.

With today's technology, I undoubtedly would have succeeded should I have taken my research any farther, but I did not have the slightest intention of setting foot in the domain of the biological sciences. I stopped at the point of confirming the possibility. When I saw insects were creating a succession of new varieties in the fields of my natural farm, I thought it would be better to leave things up to them, and I stepped back.

I also came to the conclusion that the classification of plants, placing them in species, genus, family, and order, was not only an imposition on the plants, but of no use at all to human beings. We would be better off simply appreciating all the diverse forms nature has provided and not interfering.

With the current technology of gene exchange, it has become easy to create different varieties of fruits and vegetables. I call it "the mad course of genetic engineering." Soon humans will acquire the technology to turn animals into plants, and plants into animals. But not only is such scientific meddling unnecessary, it is dangerous. If we follow Darwin's thinking, that one form of life evolved into the millions of life-forms that exist today, then it seems justifiable to add a few new species or genera here and there. People may hope that they can create even better organisms and that these new forms will assimilate with nature's creatures, but the result will be the opposite.

By creating new organisms through biological engineering, people take the risk of throwing the natural plot of the world play into confusion. Even if we understand that the functioning of the genes of living things is determined by the way the four bases of DNA are arranged, it is optimistic to suppose that genetically engineered plants and animals will not get out of control.

There is so much in this world that we do not understand, not only about the shapes and forms of living things, but also about their temperaments and spirit. When we try something like creating new lifeforms and then turn them loose in the environment, disastrous side effects are *certain* to occur—we just do not know exactly what these side effects will be yet.

Abandoning What We Think We Know

Ten years ago I paid a visit to the niece of Albert Einstein, near Central Park in New York City. When I asked her if Einstein thought that time and space really existed, she replied that although he saw time and space as relative, he probably could not give an opinion about their reality. That might well have been his answer, if I had been talking to Einstein himself, but we will never really know.

What we *do* know is that he said that mass and matter are energy. We know that extensions of this idea led to enormous energy explosions by splitting the atom. Nuclear physicists realized that if atoms could be split, they could also be fused, and so, brushing aside Einstein's second thoughts, they created the hydrogen bomb. Einstein must still be drifting around in purgatory, burdened with responsibility for his part in this tragedy. I cannot speak for him, of course, but I imagine he would enjoy a conversation about the limitations of human knowledge.

Some years ago, Fritjof Capra, a professor of theoretical physics at the University of California who also lectures on science as a holistic discipline, visited my hillside hut. He was troubled that the current theories of subatomic particles appeared to be incomplete. There ought to be some fundamental principle, Capra said, and he wanted to express it mathematically.

In searching for this elusive fundamental principle, he had found a hint in the Taoist concept of yin and yang. He called it the science of the Tao, but he added that this alone did not solve the puzzle.

He had likened the lively dance of subatomic particles to the dance of the Indian god Shiva,* but it was difficult to know what the steps of the dance were, or the melody of the flute. I had learned about the concept of subatomic particles from him, so of course I had no words that could directly dispel his frustration.

It is one thing to think that within the constant changes of all things and phenomena there must be some corresponding fixed laws, but humans cannot seem to be satisfied until they have expressed these laws mathematically. I believe there is a limit to our ability to know nature with human knowledge. When I mentioned this might be the source of his problem,

---

* A major Hindu deity, Shiva is often depicted as dancing on Apasmara, the demon of ignorance.

Capra countered, saying, "I've written more than ten books, but haven't you written books, too, thinking knowledge was useful?"

"It's true that I have written several books," I responded, "but you seem to have written your books believing they would be useful to other people. I've written mine with the idea that books are not useful at all. It appears that both of us, from the West and the East, are investigating nature and yearning for a return to nature, so we are able to sit together and have a meeting of the minds. But on the point of affirming or negating human knowledge, we seem to be moving in opposite directions, so we probably will not arrive at the same place in the end."

In the end, it will require some courage and perhaps a leap of faith for people to abandon what they think they know.

何も無いった
すべてがある

CHAPTER THREE

Healing a World In Crisis

I READ ABOUT A PROFESSOR recently who did a basic study of the deserts of Iran and Iraq. I do not remember his name anymore, but he came to the conclusion that it would be better to leave the deserts as they are, and refrain from intervening. There is also a theory that it would be better to let Africa's deserts and the resulting public health problems take their "natural" course.

When considering these issues, we should begin by asking ourselves what is normal and what is abnormal. If a desert is naturally occurring, then it is better not to interfere, but if it exists as an abnormal condition, then we have no choice but to help restore it to health. It is clear that in many desert areas nature has been laid to waste and food has become scarce as a result of human activity. Now we must pay the price and take responsibility by repairing the damage.

If we investigate scientifically what is right or wrong, healthy or ill about the earth, and also the health and illness of human beings, we may seem to understand, but there are no absolute standards for making such judgments. Better to consider all things from the beginning with an open mind.

Restoring the Earth and Its People

With our current system of observation and judgment, human beings cannot decide whether the deserts are a kind of cancer making the earth sick, or a phenomenon of self-cleaning—a change by which the earth achieves balance. People see the population increase in Africa, China, and India as tragic, but who is it that has brought about the disappearance of vegetation and the scarcity of food there?

In the past, present, and future, the true disposition of nature is toward abundance for human beings and for all species. Therefore, the question should not be "Why are there too many people?" but rather, "Who has created the scarcity into which they are born?" And then, finally, "How can we heal the earth so it can support future generations?" It is too simplistic to begin and end the conversation with a limited view of overpopulation. Better to ask: Why must people suffer so? And have we done all we can to alleviate the pain of the earth and the pain of the human race?

It is important to reflect on what has happened historically in regard to agriculture and medicine. We have seen huge advances in modern medicine, but there is little value in the advancement of medicine if the number of sick people continues to increase. It is the same with modern agriculture. How can we congratulate ourselves on the advances in modern agriculture, including greatly increased production, if the rate of starvation, scarcity, depletion, and disease increases even more rapidly?

In Nature, There Are No Beneficial or Harmful Insects

During the years I have watched the development of my natural farm, I have seen little damage done to fruits, vegetables, or grains. The crops have grown vigorously and lived natural lives without withering and dying prematurely. That does not mean it is pest- or insect-free. If you looked closely, you would observe many insects on the fruit trees and many diseased leaves. The damage they cause makes up no more than 5 percent, but that amount must be allocated to provide food for birds and insects and to thin out the weakest individuals.

Plants, people, butterflies, and dragonflies appear to be separate, individual living things, yet each is an equal and important participant in nature. They share the same mind and life spirit. They form a single living organism. To speak of creatures as beneficial insects, harmful insects, pathogenic bacteria, or troublesome birds is like saying the right hand is good and the left hand is bad. Nature is an endless cycle, in which all things participate in the same dance of life and death, living together and dying together.

Eastern and Western Medicine

In Western medicine, the body is first examined to determine which parts are ailing, and then an attempt is made to heal the ailment. That is, doctors use a localized, external treatment of symptoms. If you have a pain in your head, doctors will order a CT scan, analyze the results, surgically remove the "abnormal" portion, and try to repair the area as best as possible.

In Eastern medicine, doctors start by looking at the eyes and skin coloring, listening to what the person says, and checking the patient's complete mental and physical health. The main objective is to find what constitutes the overall *health* of the individual.

It is thought that in the end, both methods will be effective in healing, but in fact, they move in opposite directions. They can also be viewed as two poles, one with the goal of healing sickness and the other with the goal of maintaining health.

When the specialized Western medicinal approach is used, the question of what gives life and health to the whole body and mind is put off. In other words, modern Western medicine puts the human body ahead of the human spirit. This separation is a starting point for emotional anxiety among people today.

Eastern medicine, on the other hand, sees a person's natural form and the degree of health of the mind and body, and asks how to *preserve* that health. What is considered the healthy body and spirit must be based on the natural form. But in contemporary society it is becoming increasingly difficult to maintain that natural form, or even remember what it originally was, since people are increasingly living in their minds and disconnected from their bodies. To find what the natural form is for human beings, it is necessary to consider what the appropriate relationship should be between people and nature and how they should live to embody that relationship.

Recently there has been a big hubbub over the question of brain death. This is a confusing issue that involves the question of the biological life and death of human beings, as well as moral questions. It is made even more complicated by the involvement of religious views of life and death. Physicians, by being overly attached to the importance of maintaining life at all cost, often try to extend biological life even if there is no joy and no hope. They blur the boundary between life and death, departing from the realm of science.

Physicians and nurses must be guides to life. They must not specialize in simply healing sickness and giving advice on pain and medication. This means there will be times when they must give people the comfort of living truthfully and simply oversee parting and death. We could say that is the ultimate, most humane medical treatment.

Regarding questions of life and death, I think people would be better off observing how the cycles of life and death occur in nature. Imagine a meadow full of wildflowers and sweet clover with bees and a few spotted fawns grazing in sunlight. Imagine the cycle of seasons—the rhythms of growth and decay, the endless beauty. We can never understand the wonderful ways of this world, but is it not enough to simply enjoy our time here and be grateful? In the end it is love, really, that sustains our spirit. Life without love, life without joy—like a barren meadow attracting no wildlife—leads to an unpleasant environment, a sickly body, and an unhappy existence.

When I asked a Japanese youth where he found happiness, he said, "I'm happy if my life is filled with fine food and clothing, a nice place to live, a car, leisure time, and foreign travel." A young fellow from Nepal responded to the same question in this way: "From *The One-Straw Revolution* I learned that true joy comes from nature, and that we can find it by giving up our attachments." One was trying to find joy in the midst of human society, the other in the midst of nature. One was hooked on materialism, while the other was seeking to be healed.

In the desert, you can hear the sound of the wind and the sand. It is a sad, dry sound, a whispering, a kind of mournful music. The sorrowful braying of a donkey I heard in the African savanna still lingers in my ears. Wailing and squealing, it was like the cry of a child on the verge of death. The desert is also yearning to be healed.

I feel that I saw the essence of medical treatment in a hospital, if you could call it that, in a desert camp for thousands of Ethiopian refugees. Palm leaves had been placed atop several spindly poles, providing a little shade, and that was the hospital.

There was a yardstick and a scale. A child was considered ill when his or her height was too tall in proportion to the child's weight, and the patient would be given a cup of milk containing a drop of nutritional supplement. Every morning two or three hundred people gathered, including the children and the relatives attending them, but only about twenty or thirty children were deemed as sick. The goal was to be one of those who received a cup of milk that day. The children who received no milk cried and whimpered. They did not cry because they were sick, but because they had been examined and judged to be healthy.

The caring presence of the nurses at this hospital seemed to give people the courage to live. The eyes of the children jumping and playing about the area were beautiful and shining. These children—living in a remote community with no writing and no money—were innocent and openhearted. During my time there, I planted vegetable seeds in the gravel around the hospital with them. Of course, the children understood quite well how wonderful it would be if the area turned a rich green, and vegetables grew up beneath banana and papaya trees, so they gleefully scattered the seeds far and wide.

Gradually I came to realize that the process of saving the desert of the human heart and revegetating the actual desert is actually the same thing.

The Fear of Death

The fear of death, I think, is not so much a fear of the death of the body as it is a fear of the loss of the attachment to wealth and fame, and to the other worldly desires that are a part of everyday life. The degree of one's fear of death is generally proportional to the depth of one's worldly attachments and passions.

So how can we die peacefully if we do not resolve our attachments? The content of these attachments, of course, is nothing more than illusion. It is the same as when a person, believing he possesses a treasure in gold, silver, and jewels, opens the box to find only worthless bits of glass and rubble.

I have said that material things have no intrinsic value. It simply appears that they have value because people have created the conditions in which they *seem* to be valuable. Change the conditions and the value is lost. Value is born and disappears according to the whims of the times.

There is nothing for people to gain and nothing for them to lose. As long as people lived according to natural law, they could die peacefully at any time like withering grasses.

If a person dies naturally, then not only is that person at ease, but the minds of those around him are at peace, and there will be no regrets in the future. Ultimately, the one that announces the coming of death and delivers the final words is not a priest or a physician, but nature. The only thing for people to decide is how they can best achieve a death that complies with nature's will.*

---

* Mr. Fukuoka passed away on August 16, 2008, at the age of ninety-five. He had been in poor health since the autumn of 2007. In early August, he asked his physician to discontinue treatments. He passed away shortly after that, peacefully, at his home, during the Obon festival. Obon is a midsummer holiday in Japan when the ancestors visit the living for three days. It is a happy time. Villagers tend the graves, and children run and play together. On the third night, the ancestors return to a send-off of songs, dances, and fireworks. Mr. Fukuoka died on the third night of Obon.

The Question of Spirit

People have concluded that the life and death of other living things in nature is the life and death of the physical body, but with human beings there is also the question of whether life ends with death or continues after it. People so agonize over the many ideas on this subject—whether people's souls continue after death, whether there is another world where spirits go after death, whether people are born again—that they can hardly manage to simply die.

We may think we understand when and where our conscious mind originated, but actually, we do not. So what is the reality of the thing imagined to be a spirit or soul? Even if we say it is the mental activity that occurs in the brain, that does not illuminate its true character.

The way we can elucidate the true nature of the mind is to consider it from the standpoint of *mu*, the awareness people have before they become aware of themselves. It is the original mind before Descartes's "I think." The "I" Descartes referred to is nothing more than the ego. It is not the pure, spotless, transcendental mind.

The ultimate goal of the Western philosophers, who are exploring the world of the individual self, and the religious people of the East, who are seeking the transcendent self, is to elucidate the *original* mind that mysteriously occurs as part of existence itself. It is only through nature that we can see this original mind.

Anyway, none of these ideas—life, death, spirit, the soul—escapes the framework of relative thought. They are nothing more than abstract notions built up of judgments and circular reasoning based on human thinking. People have created a world of ghosts called the hereafter. But no matter how much humans search for freedom from the fear of not knowing, in the end, they should just return to the reality of nature and live their lives in peace.

The Money-Sucking Octopus Economy

The first thing I wondered about when I heard the news of the collapse of the Soviet Union was what would happen to the economies of the capitalist countries in years to come. The fundamental doctrines of capitalism and communism differ, of course, in that capitalism concentrates on production and consumption based on free competition, while communism emphasizes production and distribution on an equal and impartial basis. Contrary to popular belief, however, freedom and equality cannot coexist completely separate from imprisonment and inequality. Even if we speak of the freedom of capitalism, one cannot willfully act with *unlimited* freedom, and not everything can be distributed equally, as communism suggests. Freedom and equality exist in the mutual relationship of warp and weft,* inseparable from their shadows and from each other. There is little difference in the content of the two, when it comes down to it.

Even if our goal is to protect forests, revegetate the desert, and revolutionize agriculture, if we do

---

* These are the threads that make up woven fabric. The warp threads run lengthwise, crossed at right angles by the weft threads.

not resolve the fundamental problems of economics and people's way of living, we will not be able to accomplish anything.

I have often said that value does not lie in material goods themselves, but when people create the conditions that make them seem necessary, their value increases. The capitalist system is based on the notion of ever-increasing production and consumption of material goods, and therefore, in the modern economy, people's value or worth comes to be determined by their possessions. But if people create conditions and environments that do *not* make those things necessary, the things, no matter what they are, become valueless. Cars, for example, are not considered to be of value by people who are not in a hurry.

Economies that aim at production and consumption of unnecessary products are themselves meaningless. People could get along perfectly well without unnecessary goods if they lived a life in which nature provided everything—assuming, of course, that they had access to the natural world. But this has become increasingly difficult in the wake of commodity agriculture and the global dominance of agribusiness.

Indeed, one can ask of capitalism: "Why are human beings not satisfied, as are the birds, with what they can glean? Why do they earn their sustenance by the sweat of their brows and suffer so?"*

---

* This may be a reference to Matthew 6:25–34, which reads in part, "Look at the birds of the air: they neither sow nor reap nor gather into barns, and yet your Heavenly father feeds them. . . . Consider the lilies of the field, how they grow: they neither toil nor spin, and yet I tell you, even Solomon in all his glory was not arrayed like one of them."

I still remember the words of an Ethiopian tribesman who at first rejected my ideas of natural farming. "Are you asking me to become a farmer?" he asked. "To be attached to the soil and to accumulate things are the acts of a degraded person." This proud nomad's words are a perceptive criticism of modern society.

The time seems long ago when the more of something there was, the less expensive it was; when you could earn a profit simply by producing the things people needed—the age of local small-scale economies.

Even though pumpkins will support the lives of hundreds of people, when it became more profitable to deal in diamonds, which weigh only a few grams each, everyone stopped growing pumpkins. Once the distribution system came under the control of large business concerns, the whole price structure became cockeyed.

When I visited Europe, for example, I found that fruit was extremely expensive in Vienna, Austria. When I asked about this, I was told that the Italian farmers were refusing to grow fruit, so the price was high for what little fruit was available. The next day, when I went on to Italy, I saw a bulldozer destroying beautiful peach trees in an orchard south of Milan. When I asked the farmer why he was doing that, he said the people in Austria would not buy the fruit. The reason, of course, was that the price was too high in Vienna, causing demand to fall and the price paid to the Italian orchardist to drop. The farmer said he was following the orders of the local agricultural cooperative to "limit production," and so he was taking out his orchard.

The same day, a French newspaper published a photograph of French farmers, at the border with

Italy, overturning five or six trucks loaded with grapes in order to prevent their importation. Consumers in the French cities at the time were buying imported fruit and wine at high prices, undercutting their own farmers' understandable concern about the low prices they were getting for their produce.

This sort of thing happens because the commercial firms that stand in the position of middlemen can manipulate prices according to the information they release. If they tell consumers that prices are high because the supply of fruit is low, and tell the producers that sales are poor, then everything goes well for the middleman broker because he has control over the cash flow. Under this system no one knows the truth. Those businessmen and financiers who know and control information about the true production and cost figures determine the prices, always to their own advantage.

I call this the money-sucking octopus economy. At the center are politicians and the military-industrial-government complex (the heart of the octopus), who have centralized authority. The octopus's eight legs are the means to serve that center, which are: (1) maintenance of the transportation network, including road, rail, and air transportation; (2) control of agencies administering transportation; (3) supervision of communications; (4) establishment of an economic information network; (5) education and administrative advising; (6) control of financial institutions; (7) control of information; and (8) control of citizens' personal computers and registration.

Everything is pulled to the center with these eight legs. Although this action is carried out under the name of stimulating the regional economy of outlying areas, or maintaining regional culture, the wealth eventually accumulates in the center. The towel the octopus has tied around its head, like a sushi maker, is a ring of money, and this money, like a magnet, draws more money through its eight legs. Money attracts more money, and it goes on and on.

And what is this wealth being used for? It is used for establishing more centralized authority and strengthening armaments—more fuel for the gut of the octopus. This will lead to national enrichment and military strength—which, if allowed to escalate unchecked, results in the mad ambition to control the world. But pride goes before a fall, and in the end the octopus will either be hauled up by the master fisherman or will eat its own legs to spite itself.

This tragic dance of the money-sucking octopus is performed on the backs of the common people and the farmers. In the end, the octopus, with its legs waving wildly in every direction, is nothing more than the human comedy.

This reminds me of the time I visited Lumbini, Nepal, the birthplace of Sakyamuni Buddha. As I rested in the thick morning mist beneath a ficus tree, some local farmers appeared in twos and threes, and walked around the pond in front of me. They were turning prayer wheels and chanting, and then they disappeared. Time stopped for me. In that moment, I thought I heard the voice of the Buddha.

On my journey home that day, a person involved with maintaining sacred Buddhist sites showed me a proposal made by a Japanese architect for turning that place into a tourist spot. I was shocked. The plan was to connect the pond with a series of canals so people could worship from pleasure boats. In the center of the park, models of the great temples, churches, and shrines of the world would be built and used as hotels.

The idea was to make visiting sacred sites more convenient, shortening the time required to grasp the mind of the Buddha. Do I even have to say that it is impossible to capture the true form of the Buddha with religious theme parks and revolving lanterns?

Although the government prides itself on the fact that Japan has become an economic power, and a majority of the people consider themselves to be members of the middle class, everyone there can sense that the current prosperity and an upcoming economic crisis are sitting back-to-back.

Not long ago, more than 80 percent of the Japanese people were farmers, and only a small fraction of the people were in trades or industrial production. Now it is reversed. The primary farming industries account for barely 5 percent of the population, and even the secondary trade and manufacturing industries have been surpassed by the tertiary consumer service industries. If a typhoon of economic depression should arise, this structure will certainly collapse.

The same agrarian landscape that could be seen in Japan until just recently still exists in the farming villages of many parts of Asia, Africa, and India. In fact, one of the most common farming tools still used in many Indian villages today is the water buffalo cart. I sensed great pride in the words of a farmer who boasted to me that the design of this cart had not been changed or improved upon in the last three thousand years.

At the foundation of the so-called underdeveloped countries is a proud agrarian ethic. If these local economies were to imitate the developed countries, with their model of concentrated power and resources, the common people would be demeaned even as the country's profiteers temporarily prospered. These dignified farmers I have met see the skyscrapers of the developed nations as the tombstones of the human race. I am reminded of a Thai folksong that goes:

> *There is rice in the fields*
> *There are fish in the water*
> *The peddler shouts his wares*
> *You can buy what you want*
> *We have sown the seeds*
> *Hurry, give them water*
> *If you don't, they will die*
> *Tonight the moon is full.*

This song celebrates the belief that the greatest joy may be found among the farmers living with only the bare essentials.

The Illusion of the Law of Causality

Natural scientists have found that if you chill a cup of water, it will turn to ice, and if you heat the ice, it will turn back into the water. In the repetition of such experiments, they have seen that there is a cause and a result in the changes of matter.

When seawater is heated by the sun, it turns to vapor, rises in the sky to form clouds, turns to rain, falls on the earth, flows down in streams and rivers, and returns to the sea. With this cycle, meteorologists discern the cause of rain and clouds and think they have grasped the true nature of the water. But they do not understand the fundamental cause that explains *why* there is water on the earth and why clouds float in the sky.

When natural scientists set up measures to counter desertification, they begin by investigating its causes and the apparent results. They conduct studies of the desert environment, the climate, the soil, and the ecology of the living organisms. Then they create a plan for reforestation. In other words, as with Western medicine, they devise a swift, localized treatment of the symptoms. But the causes they base their solutions upon are not the fundamental causes. Their countermeasures serve not to heal, but rather magnify the scope of the problem.

Let me talk for a moment about my own experience with the pine forests in Japan. Lovely green pine trees growing near white sandy beaches have long been a representative landscape of the Japanese islands, but in the mid-1970s the pine trees began dying left and right. In a short time, the beautiful pines covering the hills have all but disappeared in many parts of Japan. The Regional Office of Forestry determined that the source of the damage was a nematode carried by a long-horned beetle. During the past ten years they have conducted widespread aerial chemical spraying in an attempt to exterminate the beetle.

My own village is in an area of red pine forests that nourishes the matsutake mushroom, a mycorrhizal fungus* highly prized for its flavor when cooked. I was unable to sit by and watch as the large green trees around my farm suddenly died, one after the other. Also, I had my doubts about the way in which the Office of Forestry had determined that the beetle and the nematode were the cause, so I put my past experience in plant pathology to use and spent more than three years doing research in one of my hillside huts.

The Office of Forestry's theory was that when the beetles laid their eggs in the tops of the pines, the nematodes living in association with the beetles invaded the pines, entered the tree's vascular system, and multiplied, blocking the passage of water and nutrients within the trunk and branches. This was said to cause the pines to suddenly wither and die. My experiments, however, showed a completely different result.

First of all, healthy pines are not likely to die, even when inoculated with nematodes. Second, I did not find the filamentous fungi that feed the nematodes (according to the Office of Forestry, nonspecific blue and black molds) present in the trunks of healthy pines, and the nematodes cannot live on pine sap alone. Then, when I studied the trunks of pines that had begun to show signs of dying, I discovered three or four *different* types of pathogenic fungi (*eumycetes*) that had not even been mentioned. These organisms are thought to have been introduced with imported

---

* A fungus that lives partly in a plant's roots and partly in the soil. It is a relationship that benefits both the host plant and the fungus.

lumber. But even when I inoculated pines with the hyphae that these fungi produced, they had little effect on the overall health of the tree.

Most fascinating from my research was the discovery that the pines first showed physiological abnormalities only after the mycorrhizal matsutake fungi were no longer present. The pine and matsutake fungi have a symbiotic relationship. The matsutake penetrates and breaks down the minerals in the soil, absorbing minute amounts of their nutrients, which it then supplies to the pine. In other words, the decline of the pine trees and the sudden decline in the matsutake fungus, which is occurring throughout Japan, seemed to be directly related.

What has given rise to the change among the communities of microorganisms in the soil of the pine forests in Japan? It is widely known that the soil of the pine forests is becoming more and more acidic, and that alone may account for the changes among the microorganisms. I believe that acid rain is the source of the acidity, but I cannot say so with certainty.

I have not yet reached a conclusion, but in relation to the cause, the results of my research point in an entirely different direction than does the theory of the Office of Forestry. The source of the problem is not a troublesome nematode, but rather that the matsutake fungi are dying, and as a result the pines have grown weak. Filamentous fungi have invaded their trunks, and finally, nematodes that feed on those fungi have invaded the pines. The nematodes and beetles are not the original culprits. They are doing nothing more than clearing away the corpses of the dead and dying trees.

At the very least, they are only accessories to the crime, while the ringleader lies underground. But even if my theory is correct, if smog and acid rain turn out to be the source of the problem, then we are nearly back where we started from.

Of course, this research was done in my crude hut in the hills, so there is room for error, but the point is, what the world sees as cause and effect can be deceptive. Although I speak of the cycle of cause and effect, no one *really* knows what is happening. Still, the Office of Forestry goes out and sprays insecticide all over the forests. Who knows what unforeseen and potentially more serious environmental disaster that may lead to?

The Current Approach of Desertification Countermeasures

When people see that rain does not fall and there is no water in the desert, the first thing they think of is to build a dam to store river water. Then they build waterways and irrigation canals. They think that in order to use water efficiently, it is best to make straight waterways so that the water will flow faster. The Aswan Dam on the Nile has already been built and soon there will be similar large dams on the Yellow River in China, and the Narmada River in India. These may serve as expedient, short-term measures, but they will turn into long-range hundred-year mistakes.

The main reason water is disappearing from the rivers is that rain has stopped falling. The first step we must take in countering desertification is not to redirect the flow of rivers, but to cause rain to fall again. This involves revegetation.

Trying to revegetate the deserts by using the scarce water remaining in the rivers is putting the cart before the horse. No, we must first revegetate vast stretches of desert at one time, thereby causing rain clouds to rise from the earth.

There is now a plan to construct more than two hundred dams along the course of the Narmada River, the second holiest river in India after the Ganges. But when the dammed waters rise, it will submerge the forests, destroying the livelihoods of the people who live there, and millions will be driven to the deserts surrounding the river. This influx of people will put greater pressure on scant desert resources, and with the loss of the forests, the deserts will grow even larger.*

The Indian government needs to decide whether it is better to carry out their hundred-year national plan of building hydroelectric power plants, or to revegetate the desert and bring the earth back to life.

It is also important to note that even if there are no rivers on the surface of the desert, there is water below the surface. In Saudi Arabia and in the desert area east of the Rocky Mountains in North America, water is pumped from aquifers hundreds of feet below the earth's surface. In the United States, the water is pumped to pivot farms, where it is dispersed by enormous sprinklers that form circles up to half a mile in diameter. Seeing the green circles these farms make in the desert, one can see the power of modern technology. But what

---

* Sadly, since the time Mr. Fukuoka wrote this in 1992, this has come to pass.

will happen when these sprinklers have pumped all the remaining underground water that has been filtering down from the forests of the Rocky Mountains and collecting for tens of thousands of years?

When water is scarce, people think about economizing. With drip irrigation, plastic pipes are laid in the desert in an effort to use the least amount of water to the greatest effect. A Japanese university has been carrying out experiments using this sort of irrigation in Mexico, and it has been employed in Israel for some time. Of course, this method is effective as a localized remedy, but in view of the overall materials and energy needed, it is questionable whether or not drip irrigation can be adopted as a practical, long-term solution.

One experiment being carried out in Egypt plows super-absorbent resins into the soil to increase its ability to retain water. Other experiments focus on various types of water-retaining materials to use in place of humus, but these measures are also short-term expedients.

In China, the government has begun a massive effort to halt the expansion of China's great Taklamakan Desert. As in Sudan and Tanzania, they are using satellites and airplanes to conduct remote surveys of the current scale of desertification. They analyze underground water, soil salinity, and other conditions by boring holes to great depths, quantifying the results, making computer simulations, and testing the tolerance of plants raised in the harsh desert environment. But every plan for revegetation created in this way has failed. Then the governments in charge of the studies say they need to go back and reanalyze the data—which will, of course, cost more time and money.

Sure, you can create excellent fields in an arid area if you pump water from underground and sprinkle it on the desert, as they do on American pivot farms. But because irrigation water that has been applied in this way quickly evaporates, the salt is precipitated out, and builds up in the surface layers of the soil. To prevent this salinization, the salts from the irrigation water are drained into rivers or nearby "dump sites," causing toxic conditions there instead.*

Methods such as those used in Saudi Arabia, Israel, and other places of filtering seawater with a synthetic resin membrane, removing the salt to turn the seawater into fresh water, circulating it, and thus creating farms in the middle of the desert, are also nothing more than short-term measures that require tremendous amounts of energy.

Some people think that to increase the vegetation on earth, it is best to plant trees that mature quickly. Today various types of fast-growing trees such as eucalyptus are being planted all over the world. These trees, however, typically require a lot of water when they are young, so a strenuous effort would have to be made to water the trees properly. When trees are watered only

---

* A well-known and tragic example of this occurred on the arid west side of the San Joaquin Valley in California. Water was pumped from an aquifer to flush salts from the soil in order to grow cotton and other crops. The salt-laden irrigation water was then drained into the Kesterson Wildlife Refuge, causing a toxic accumulation of salts and selenium. Selenium is naturally occurring and is not harmful in small amounts, but at that high concentration, it was deadly. In 1983, massive numbers of fish were found dead, as were thousands of dead and deformed birds. The practice was discontinued in 1985, but high concentrations of selenium, lead, boron, and other toxins remain.

to a shallow depth, the soil becomes compacted. Then the water cannot percolate deeply into the ground, roots cannot extend, and in the end you might as well have poured water on heated rocks. For this reason many of the trees wither and die.

Desertification caused by sheep, cows, and goats is also a serious problem. Not only do they eat the sparse vegetation, but they also eat the trees that people plant with the goal of restoration. It appears that we must decide how to control the number of domestic animals and how to give them proper management—a system that encourages healing and regenerating the soil upon which they graze—at the same time that we deal with the problem of the world's human population.

When they see that the food supply is insufficient, people hurry to cut down trees and try to grow crops as quickly as possible, often in burnt-over fields. Overall, however, vegetation is decreasing faster than it is regenerating, and desertification proceeds at an accelerating rate.

Scientific revegetation measures often consider only one route to healing the desert. As it is now, the various scattered, localized efforts to halt desertification end up as half-baked measures administered by government officials. The problem is that the water, soil, and plants are considered separately, with each being advanced by a separate department. A permanent solution will never come about in this way.

For the past fifty years or so, I have grown crops without tilling the soil and without using fertilizers or agricultural chemicals. I have done practically nothing, and the soil in my fields has become the best in my village. I simply scattered seeds in clay pellets, covered them with straw, and grew a healthy ground cover including white clover and vetch. I supplied nature with the tools, and then I relied on nature's disposition toward fertility. Although the climate and other conditions are different, I believe that this basic method will also work in revegetating the deserts.

[handwritten annotation: Has it been tried?]

本朱
東西
高底
遠近
苦楽
愛憎 等

砂漠て
蛭を育つ
人

CHAPTER FOUR
===

Global Desertification

ALTHOUGH THE SURFACE of the ground in Europe and the United States appears to be covered with a lovely green, it is only the imitation green of a managed landscape. Beneath the surface, the soil is becoming depleted due to the mistaken agricultural practices of the last two thousand years.

Much of Africa is devoid of vegetation today, while just a few hundred years ago it was covered by deep forests. According to the Statistical Research Bureau in India, the vegetation there has also disappeared rapidly over the past forty-five or fifty years and now covers less than 10 percent of the land's surface. When I went to Nepal, officials lamented the fact that in the last twenty years the Himalayas have become bald, treeless mountains.

In the Philippines, on the islands of Cebu and Mindanao, there are banana plantations but no forests, and there is concern that in a few years even drinking water may be in short supply. In Thailand, Malaysia, and Indonesia, as farming methods that protected nature have been swallowed up by the wave of modern civilization, the condition of the land has also deteriorated. If the deforestation of the tropical rain forests in Asia and Brazil continues at the present rate, oxygen will become scarce on earth and the joy of springtime on the planet will be replaced by the barrenness of winter.

The immediate cause of the rapid loss of vegetation has been the indiscriminate deforestation and large-scale agriculture carried out in order to support the materialistic cultures of the developed countries, but the remote cause stretches back thousands of years.

The natural world did not become a desert on its own. Both in the past and at present, human beings, with their "superior" knowledge, have been the ringleader in turning the earth and the human heart into wastelands. If we eliminate the fundamental cause of this destruction—people's knowledge and actions—nature will surely come to life again. I am not proposing to do away with human beings, but rather to change the politics and practice of our authority.

My measures for countering desertification are exactly the same as the basic natural farming method. One could refer to it as a natural farming revolution whose goal is to return the earth to the paradise it once was.

Lessons from the Landscapes of Europe and the United States

I first saw the desert and began to have an interest in it the summer I flew to the United States for the first time, in 1979. I was expecting the American continent to be a vast, fertile green plain with lush forests, but to my amazement, it was a brown, desolate semi-desert.

I gave a talk in Sacramento, California, for the state's Department of Conservation hosted by Ms. Priscilla Grew, who was head of the department at the time. I said that the environment in California had serious problems as a result of poor agricultural practices, poor water management, overlogging, and overgrazing. These things, I told the group, are conspiring to create the "Great California Desert." After the talk, I was invited for a private conversation with Ms. Grew, a geologist, in her office on the thirteenth floor of the Resources Building.

We discussed how Japan and California were roughly the same latitude, that both the vegetation and the parent rocks in the two places were similar, and that long ago the Asian and American continents were one. The fossil record shows, for example, that vast forests of *Metasequoia** existed in both places. The mosses and lichens I saw growing in the undisturbed forests of the Sierra Nevada and the Coast Range were also the same as I observed in the virgin forests of Japan.

It was my conjecture that the desertification and climate change in California has been accelerated by mistaken agricultural methods. I suggested that deforestation and the change from the perennial bunchgrasses that once covered the plains to annuals such as foxtail and wild oats contributed to the decrease

---

* This is one of three types of redwood trees, the others being the coast redwood and the giant sequoia. The *Metasequoia*, or dawn redwood, was thought to be extinct until a few groves were discovered in southern China in 1944. It is now a popular landscape tree widely available in plant nurseries.

in rainfall.* "Rain doesn't only fall from the sky," I suggested. "It also falls up from below." The vegetation, especially trees, actually *causes* the rain to fall.

After we left her office, someone suggested that I come along and visit an interesting place nearby. That "interesting place nearby" turned out to be a hot, dry plateau in the Coast Range about one hundred miles away.

About twenty young people from several countries were somehow managing to live in this remote area on national forest land. They asked me to teach them how to use natural farming to help them make their livelihood. They did not even have proper sickles or hoes. The entire area was covered with dry grass, with not a spot of green in sight. There were only a few oak trees here and there.

In the midst of such hopeless circumstances, I was unable to sleep. Early the next morning, as I was washing my face at a small spring, I noticed that water soaking a mouse's nest had caused some weed seeds to sprout and grow a few inches tall.

I had always thought that the grass in California died because the summers are hot and dry, but I realized it was only the introduced annual grasses that gave that impression. They come up in the fall with the first rain,

---

* The grasslands of California originally consisted of perennial grasses. These plants have deep and extensive root systems and stay green all summer. When the Spanish introduced grazing sheep and cattle in the late 1700s, they also brought the seeds of annual grasses such as rye and oats. The grazing animals selectively ate the more nutritious native perennials, giving the annuals a big reproductive advantage. The native grasses were supplanted by the annuals in just a few generations, leaving the soil depleted and much drier.

set seeds, and then die by early summer. These annuals had chased out the native grasses, which remain green all summer. Grazing probably had a lot to do with it, but there were no grazing animals in the area anymore. Thinking that the green perennial plants ought to come back if we got rid of the weedy annuals, I set about doing an experiment.

After broadcasting the seeds of various Japanese vegetables amid the dried grasses and mowing them down with an improvised sickle, I brought water from the spring near the top of the hill by plastic pipe and sprinkled it fairly deeply over the area. I thought the few days until the water evaporated would tell the tale. Eventually, green began to grow among the brown grass. Of course, it was the green of the weedy foxtails. As I expected, when the water had disappeared by the end of a week's time, the grass that had sprouted up began to wither in the heat, but in its midst Japanese pumpkin, cucumbers, tomatoes, okra, daikon, and corn began to flourish. The center of the field turned into a vegetable garden. The stubborn foxtails sprouted, then withered and became mulch, and in their place, vegetables had grown up.*

---

* This technique of watering annual weeds to get them to sprout and then wither before they can set new seeds—known as premature germination—has been used by organic farmers for many years to control weeds. When the weeds grow up, they shade and cool the ground long enough for the vegetables to get off to a good start, then they act as mulch for the vegetable garden, cooling the ground and conserving moisture. When the autumn rains arrive, fewer weeds come up since they were "tricked" into germinating too soon. Mr. Fukuoka is suggesting that this technique could also be useful in broad-scale rehabilitation for establishing trees, shrubs, and perennial grasses.

We should revegetate California. We should wake up the seeds of weeds that are lying dormant during summer by giving them water, and then let them die before they can make more seeds. At the same time, it would be good if the state government would broadcast seeds of perennial grasses from the air in clay pellets. After this experiment, however, I had to press on with my travels, so I left the mountain and entrusted these hardy souls with my dream.

Later that year, I was shown around Europe by a Greek gentleman and a young Italian woman who had stayed in one of my hillside huts. The European countries are, for the most part, very careful about protecting the natural environment and maintaining the lovely vegetation. At first glance the entire area looks like a natural park, but it is only the beauty of a picture postcard. If you look closely, you will find that there are very few varieties of trees. The soil is thin, hard, and unfertile. It appeared to me that the earth in Europe had been damaged by an agriculture made up of mismanaged pastures used to produce meat for royalty, and vineyards to produce wine for church use.

Generally speaking, the farther south you go from the Netherlands, up the Rhine, and toward Italy, the more the number of trees decreases and the green color fades. In addition, much of the Alps are composed of limestone and have few large trees. The farther south you go, the higher the soil temperature, and the drier the climate. The soil becomes thinner and increasingly less fertile. My impression was that in Europe, the soil was dry and depleted just below the surface.

When people started plowing, that marked the beginning of modern European civilization. *Culture*, in its original sense means "to till the soil with a plow." When tractors were introduced, production increased, but the earth lost its vitality even more quickly. Throughout human history civilizations have been founded in areas with rich soil and other resources. After the soil was depleted as a result of cutting too many trees, overgrazing, harmful irrigation practices and plowed-field agriculture, that civilization, which had been wearing the mask of prosperity, declined and often disappeared altogether. This has happened over and over again.

From my observations in Europe and the United States, I could see how the errors of modern agriculture were damaging the earth. That strengthened my conviction that natural farming methods are the only ones capable of reversing this degradation. In order to prove this, I set my sights on Africa.

The Tragedy of Africa

I heard of a plan by some people in a private, nongovernmental organization to promote modern agricultural methods in Somalia, so I joined up with them, flying to Africa, hoping to test my natural farming method for revegetating the desert. My first surprise came as I flew over Somalia and saw the large Juba River, the waters of which flow through the semi-desert year-round. The source of this river lies in the distant mountains of Ethiopia.

As the river nears the Indian Ocean, in some areas it disappears beneath the sand. At one time there must have been many such invisible rivers in Africa because, when viewed from the air, I could almost always see one or two pools of water in the midst of this semi-desert, which the local people call "the land of thorn trees."

Also, when traveling over land, I saw large trees of unknown varieties. People told me that several hundred years ago these large trees formed a dense forest. Naturally, I tried to find out why the forest had disappeared.

From the accounts given to me by an Ethiopian elder and some Somali farmers, the main cause was the colonial agricultural policies brought in by Westerners. They introduced and exclusively grew commercial crops such as coffee, tea, sugarcane, cotton, tobacco, peanuts, and corn. Production of personal food crops was forbidden. This was done in the name of enriching the national economy.

When I went to apply for a visa from the Somalian government, I was flabbergasted when they told me that any kind of instruction that agitates the farmers and encourages them to become self-sufficient would not be welcome. If such activity went too far, they said, it would be considered treason.

Today, after two hundred years of colonial rule, seeds of the crops necessary for self-sufficiency have all but disappeared in Africa. If the seeds are gone, and the farmers are reduced to growing only cash crops, they descend from being farmers to simple laborers. They will have no chance of standing on their own feet again, and any possibility of agriculture that benefits nature will be cut off. Because the land cannot support the continuous cultivation of coffee and sugarcane, other seeds must be sown to restore the natural cycle, leading to healthy soil.

If the first cause of the African desert is errors in farming, the second is mistaken policies regarding the nomadic people. In the past, many African people were nomadic. From about twenty-three hundred years ago, they lived by roaming freely about the hills and plains with their camels and goats. Under colonial rule, however, this way of life was forbidden.

Originally there were various tribes, but no nation-states. As the Westerners conquered them and drew arbitrary boundaries to form countries, they also created large national parks, in the name of preserving the natural environment and animals. Common people were not allowed to enter the parks freely or graze animals there.

This was a deathblow to the nomadic people. On the surface, the old grazing patterns appeared to be random, but in fact they were bound by strict tribal rules. With the arrival of national parks, this tribal system broke down.

In the old days, the nomads, along with their domestic animals, would live in a green valley for a set time, say, three months. When the grass that fed the animals grew scarce, the tribe would move to another location. They would leave before the vegetation lost its ability to recover. No one would allow their animals to graze in those areas for six months or a year. When the grass had recovered and grown luxuriant again, members of another tribe would move in and start living there. This sort of practice was an unspoken arrangement that was strictly obeyed.

But once national boundaries were drawn and parks created, the nomadic peoples had to travel long distances to go around them. Because of the inconvenience, they began staying in one place for a long time. When this happened, fodder grew scarce, the trees used for fuel were all cut down, and the supply of water became exhausted. When food, clothing, and shelter are restricted, conflicts occur, and people begin struggling with one another. This condition was often used to maintain political control by those in power.

A teacher who was involved in South Africa's anti-apartheid movement came to my hillside hut in Japan once and asked to be taught the methods of natural farming. It happens that when he came I was harvesting rice with a sickle. He said he wanted to learn how to use a sickle. It seemed from his stories that most of the people in South Africa do not know how to sow seeds or to take care of plants once they have sprouted. When I heard from him several years later, he said that he was having greater success teaching people that they could achieve independence by starting a natural farm than he had by spreading the political message of the independence movement.

The Africans may appear at present to be enduring extreme poverty, but they are a very proud people. Perhaps the nomadic people who rambled at will over the hills and plains of Africa lived the same sort of life as the Zen monks who wandered as freely as the clouds. Indeed, when I saw the reverent form of a Somali youth praying toward Mecca in the brilliant red sunset, I felt I was seeing the eternal Africa.

Sowing Seeds in an African Refugee Camp

I was told that giving seeds to the nomads might be considered disrespectful because farmers are considered to be of a lower class than nomads. A young Japanese man who was traveling with me suggested that they might feel insulted and it could possibly even be dangerous, but I ignored his words. I went into a settlement of Ethiopian refugees and began handing out seeds.

At first, only the children approached me, probably because I was such a strange sight. When I gave them a handful of seeds, they clutched them so hard their nails left prints on their palms. "What is this? Is it food?" they asked. When I told them not to eat the seeds, but to plant them in the ground and water them for three days, they would shrug and say, "We can't understand your strange English." Of course they did not understand a word of Japanese.

After four or five days, about twenty or thirty children came and asked me to follow them. I found that each child had made a circular patch in the sand four or five feet in diameter in which cucumber, squash, beans, eggplant, and daikon plants had sprouted. Three of the youths started beating on a broken bucket and singing. My heart was warmed by this sight. After that, even women and the elderly joined in sowing seeds.

The speed of plant growth there is truly amazing. Everything, including Japanese fruits, flourishes. In particular, oranges, grapes, and pomegranates grow two or three times faster than I was used to. A papaya will bear about ten fruits in four to six months, while bananas set fruit in one year. The bananas ripen in a year and a half.

Modern agriculture in the desert is based on the idea that you can grow anything if you just have water. Every day they pump water up from the rivers, run it into irrigation ditches, and water the fields in succession. The large modern farms sponsored by foreign aid all follow this practice. As I had expected, all the modern farms I saw had failed. They had turned to salt fields and had been abandoned.

I advised people, on the contrary, to use as little water as possible in the desert. I encouraged them to plant acacia trees with a mixture of vegetables and grains such as deccan grass (*Echinochloa colona*) and millet. I also suggested that they include some poisonous plants that goats will not eat, and to plant trees that are effective at bringing water up from underground. Among the trees, they should sow grains and vegetables.

I also tried planting bamboo, reeds, and willows in the sand to serve as erosion barriers along the riversides. The bamboo, in particular, seemed like a good possibility. Sugarcane would also be good for shade and erosion control.

There were many fish in the river. Although the tribe's people are suffering from malnutrition, they are not accustomed to eating fish because of a religious belief that if you eat a fish or a snake, you will become one in the future. But in the end the youths gladly ate the small dried fish I had brought with me from Japan and they said it was delicious.

I talked with one tribal elder at length about his community's situation. "Rain has stopped falling in

Africa, and so we can't do anything. The earth seems to have died," he lamented.

I answered, "It may seem that earth polluted by chemical fertilizers, pesticides, and herbicides has died, but the soil of Africa is just resting. The red clay is taking a nap. If the people will work to awaken the sleeping soil, then you will be able to grow anything."

"What do we do to wake it up? Tell me scientifically," he replied.

"The problem is not that the soil is deficient in nutrients such as nitrogen, phosphorus, and potassium. The problem is that these nutrients have been absorbed by the clay and are not soluble in water, so the plant's roots are not able to absorb them. What you need are scissors for cutting the nutrients from the clay."

He laughed and said, "The only one with such handy scissors is a crab."

I responded, "The microorganisms in the soil will do it for you, without your having to work hard at all. You don't even need to know anything about microorganisms. When you sow the seeds of crops and trees, just be sure to mix in the seeds of legumes such as Egyptian clover (*Trifolium alexandrinum*) and alfalfa with them. The more partners there are, the better. As the life in the soil returns, the nutrients will become available to the plants once again." When I explained it this way, he seemed to understand.

In the desert, high temperatures from radiant heat are of greater concern than lack of water. This, too, will change if the surface of the soil is covered with vegetation. With a healthy ground cover, the soil temperature is moderated to a great degree.

月みつや
星めつり
シャンデリヤ
そんなが
いらぶいよ

鈴木さん
貴女つはふに
してそうり
私の主人が
言ふよ

CHAPTER FIVE

Revegetating the Earth Through Natural Methods

WHEN THE DIRECTOR of the Desertification Prevention Council at the United Nations told me that he thought my natural farming methods could be useful in preventing the spread of the deserts, I was surprised. Looking back, though, I realize that the techniques I developed over those many years in Japan *could*, in fact, be used effectively to counteract desertification.

People have been living on the site of my natural farm since the Stone Age, and in ancient times the area was covered with an ancient forest of at least eight species of *Metasequoia*. It is said to have been a center of local culture about a thousand years ago. The area could be compared to the Silk Road in the way the culture flourished until the earth was eventually worn down, and the vibrant culture with it. The soil of this once fertile forestland eventually eroded down to the clay subsoil.

Years ago, at the site of my natural farm, people tried planting mandarin orange trees, but the trees did not thrive, so they largely abandoned the land. That is the land I started with. Since then I have turned the soil of my family's orchard into soil as fertile as the forest soil it once was.

When I returned to my father's farm after the war, the trees in the citrus orchard were struggling. There was little vegetation on the surface, and the ground had become compacted. At first I thought the soil would improve quickly if I brought ferns and rotting tree trunks I found in the forest and buried them, but the experiment was a failure, mainly because it took too much work to bury enough material to make a significant difference. That organic farming technique might have eventually helped improve the soil, but it required a lot of labor for very little result. Like many other farming techniques, this was one that resulted in a net loss.

The start of my "do-nothing" farming method came when I decided to simply plant a haphazard mixture of fruit trees, vegetables, grains, and clover among the acacia trees growing on the hillside and then observed which plants thrived and how they got along with one another. I neither cleared the land nor tilled it, and I used no fertilizers, herbicides, or pesticides. Now, some forty-five years later, it has turned into a fruit tree jungle. I believe that the techniques I have perfected after all these years can also be applied to regreening the deserts.*

---

* Mr. Fukuoka is not saying that his farm was a desert when he started farming there, but that the soil was depleted and needed to be reinvigorated. In large part, the development of his natural farm was itself a rehabilitation project.

And so, while my journeys in later life have been inspired by my dream of revegetating the desert, unlike the typical scientist I have not tried to amass data or systematically formulate measures for preventing desertification. Instead, my desert prevention measures are strictly intuitive and based on observation. I arrived at them by using a deductive method. In other words, I started with the recognition that the causes of desertification in most areas are misguided human knowledge and action. If we eliminated them, I believed that nature would certainly heal itself.

Unfortunately, there are some places where the devastation is so bad that nature will have a difficult time recovering on its own—it lacks even the seeds that would form the basis of its recovery. The only work for people to do in such places is to gather the seeds and microorganisms nature needs and sow them there.

A good example is the natural farm of Ms. Aveliw, a woman who works for the Magsaysay Foundation in the Philippines. She read *The One-Straw Revolution*, did practical research and observation on her property for almost ten years, then set up her natural farm in four years mainly by scattering seeds and planting trees. She has created a true paradise. There is an assortment of fruit trees such as banana, papaya, guava, and durian, and some coffee trees. Beneath them is a thick ground cover of perennials and green manure. Orchids bloom everywhere, birds fly about, and fish swim in the ponds.

The soil of the Philippine islands is rather poor, and, as a result of reckless deforestation, you cannot find anything like a true tropical rain forest anywhere. So how was she able to develop this abundant jungle of fruit trees in so few years? The secret element was the harmony that existed between Ms. Aveliw and nature.

If we list the things necessary for plants to grow, then sunlight, nutrients, water, and air are sufficient to create a paradise. All the elements are created by nature itself. Even without instruments, nature is capable of performing a splendid symphony.

If you believe in intuitive insight, the road will open on its own accord. If you believe that the Philippines were originally a paradise, and sow seeds there, nature will create a forest of abundance and beauty. I was deeply moved by Ms. Aveliw's example.

Agricultural "Production" Is Actually Deduction

Just as the vegetation on the plains of Europe and America has become scant as a result of deforestation, the degeneration of the mountain plateaus and the flatlands has become pronounced in the Philippines, Thailand, and India. The Asian countries where clear-cutting is being carried out are even worse. In those places, the deep green of the jungle forests has all but disappeared. Mountains are stripped bare, with only a few small trees left behind. The trees in the bottomlands and near waterways have been cut down as land is cleared for agriculture.

It is not generally noticed, but the destruction of forests is gradually having an effect on paddy agriculture. While there is still some fertile soil in the rice-growing regions of the Philippines, Thailand, and

India, I am worried about how long the productivity of these regions can continue. My immediate concern is that unforeseen changes are occurring in the communities of plants and soil microorganisms as a result of using chemicals.

There are also instances in which so-called improved varieties of pasture grass have turned into weeds that are difficult to control. Mechanization and the heavy use of agricultural chemicals are not only ruining the land, but also causing the ruin of farming communities around the world.

In addition, while modern agriculture appears to be increasing yields, net productivity is actually decreasing. If we compare the energy required to produce a crop of rice and barley with the energy harvested in the food itself, we find a disturbing trend. Fifty years ago in the United States, each calorie of energy invested to grow rice resulted in a yield of about two calories of grain. Thirty or forty years ago the two figures became equal, and now, the investment of two calories of energy produces only one calorie of grain. This is largely because of the shift from using such things as hand labor, draft animals, and cover crops to using machinery and chemicals, which also require factories to create the tractors and chemicals, and mining and drilling to produce the raw materials and fossil fuels.

The soil in Japan is quite fertile, and for a long time traditional Japanese agriculture maintained harvests of more than two calories of rice for each calorie of energy invested. As mechanization was introduced, however, increased harvests became the overriding goal, and efficiency declined sharply. Now, as in the United

States, the energy produced is only half that invested. In other words, we have *deduction* rather than production.

So in terms of energy production, modern petroleum-based farming is not producing anything at all. Actually, it is "producing" a loss. The more that is produced, the more of the earth's resources are being eaten up. In addition, it creates pollution and destroys the soil. The apparent increase in food production is also subsidized by our rapid depletion of the soil's organic matter. We are simply squandering this gift of stored solar energy.

We are the octopus congratulating itself for becoming fat by eating its own legs. High-tech farming methods give the illusion that even if the earth loses its fertility, even if there is no rich soil, food can still be reliably produced. But if petroleum becomes even a little scarce, food production will take a sharp and immediate drop. When we have reached the point at which obtaining one unit of food energy requires three to four times as much energy input, how will the human race be able to maintain its food supply?

There is no technology for increased food production that uses more energy than high technology. Therefore, whoever controls petroleum can control the world's food supply. I find this situation really disturbing.

Commercial Feedlots Will Destroy the Land, Cultured Fish the Sea

The modern livestock and fish industries also have fundamental problems. At their inception, it was casually thought that by raising more domestic animals and cultivating more fish, our diets would be enriched. But by using mass-production techniques, the meat and fish industries severely pollute the earth and the sea.

When we look at calories in terms of production and consumption, if people want to eat eggs and milk, they have to go to twice as much trouble as if they ate grains and vegetables; to eat meat raised on commercial feedlots, they must expend seven times the effort. That is because the contemporary methods for raising cattle are a waste of energy. Just as with industrially produced plant crops, the industrialized meat industry should not be called a production activity at all.

Our cattle in Japan are fed on corn grown in the United States, which is brought all the way here by ship. The cows are raised in small stalls and waited on hand and foot by the farmers who feed them. The cattle never get to graze on pasture grass. It is the same as the confined animal feedlot operations in the United States—both rely on the massive production of corn, to name only one problem. Another is the sorry waste of useful animal manure that could otherwise be scattered evenly across a grassy pasture to improve the condition of the soil.

The same is true for the modern fish-producing industries. Their methods are destroying the mangrove forests along the Asian coastline, and destroying the seas that once were abundant fishing grounds. These "producers" typically use about ten pounds of small fish as food to raise one pound of high-class seafood. Then they congratulate themselves on how efficient their operation is.

In order to protect natural fisheries, we ought to go back to catching fish by hand. Fish will not become more abundant if we continue to develop and apply new technologies for cultivating shrimp, sea bream, and eel. Eventually such methods will lead to the collapse of the modern fishing industry and the seas themselves.

In what I would consider to be an ideal situation for raising cows and other farm animals, the flowers of clover and vegetables would bloom in an orchard of trees laden with fruit and nuts. Bees would fly among the barley and wild mustard that had been sown here and there and later reseeded by themselves. Chickens and rabbits would forage on whatever they could find. Ducks and geese would paddle about in the ponds with fish swimming below. At the foot of the hills and in the valleys, pigs and wild boars would fatten themselves on worms and crayfish, while goats would occasionally peek out from among the trees in the woods.

Scenes like this can still be found in the poor villages of some countries not yet swallowed up by modern civilization. The real question is whether we see this way of life as uneconomical and primitive, or as a superb organic community in which people, animals, and nature are one. A pleasant living environment for animals is also a utopia for human beings.

Sowing Seeds in the Desert

Without questioning whether they were native or non-native, I would mix the seeds of all plants—forest trees, fruit trees, perennials, vegetables, grasses, and legumes—as well as ferns, mosses, and lichens, and sow them all at once across the desert.* I would even include fungi, bacteria, and other soil microorganisms. If it were possible, I would scatter black forest soil as well. Fertile soil is a valuable trove of microorganisms and their spores. In extreme deserts, this would be the most economical means of reintroducing them. It would also be good to include seeds of the native plants that grew before desertification occurred in California and India, heat-resistant plants in Africa and Thailand, and salt-resistant plants in Somalia.

I would encase these seeds and microorganisms in clay pellets˟ and sow them on a broad scale. These pellets protect the seeds from animals and insects and maintain the water necessary for germination. They also serve as survival capsules, providing nutrients after germination. To make the pellets hard, I use lime, bittern, seaweed paste and other binding compounds. To make them resistant to insects, I mix in herbs such as derris (*Derris elliptica*), Japanese star anise (*Illicium anisatum*), Japanese andromeda (*Pieris* spp.), lacquer tree (*Rhus verniciflua*), and Japanese bead tree (Chinaberry, *Melia azedarach*).

---

* The method Mr. Fukuoka describes in this section is meant for rehabilitating damaged landscapes, especially deserts. He does not recommend using this technique in pristine areas like the Siskiyou Mountains in southern Oregon and northern California, or the Olympic Mountains in northwestern Washington, where the soil and native plant and animal communities are still largely intact.

˟ The complete method for making clay seed pellets is given in the appendix.

Then I would broadcast the seeds in the clay pellets and wait for rain. If there were a squall or thunderstorm, the seeds would germinate and grow very quickly. If a large area became green, even temporarily, it would protect against radiant heat, and the soil temperature would drop. Cooling the soil is an important step toward success.

I would do this resigned to the fact that if a drought followed the rain, a majority of the plants might die. But even if most died, *some* of the plants, especially those that could withstand heat and could thrive with little water, would survive. In the shade of these plants, grasses, shrubs, and trees would grow up here and there. Then even if we left the place alone after that, green would summon more green, insects would come, birds would come, small animals would come, and they would all scatter seeds. If one tree grew, it would act as a pump to bring up underground water. The mist transpired from its leaves would act as both a sprinkler and a fan. When various kinds of plants, large and small, grew up, their geometric increase would be greater than one can imagine.

I have said that it would be good to include seeds and spores from all over the world in the mix. There are several reasons why. The world has become smaller. People travel freely everywhere, carrying seeds and microorganisms with them. The reality is that the plants and animals of the world have already been thoroughly mixed together, and that will continue. Plant quarantine systems have rapidly become obsolete. It seems to me that the time has come to abolish these quarantine regulations entirely.

We cannot simply put things back the way they once were. Too much has happened. Conditions are far different today from what they were just one hundred years ago. Soil has eroded and become drier due to agriculture, overgrazing, and cutting too many trees. Plant communities and the balance of microorganisms have been altered beyond recognition by plowing and agricultural chemicals. Animals and plants are becoming extinct from the elimination of their habitat. The seas are becoming more acidic, and even the climate is changing. Even if we did go to the trouble of putting back the plants that were native to a certain place, there is no guarantee that they would thrive there anymore.

My idea is entirely different. I think we should mix all the species together and scatter them worldwide, completely doing away with their uneven distribution. This would give nature a full palette to work with as it establishes a new balance given the current conditions. I call this the Second Genesis.

Creating Greenbelts

My measures for revegetating the desert are essentially the same as the ones I used in establishing my natural farm. The fundamental concept of a natural farm, as I have described, begins with intuitively grasping nature's original form, where many varieties of plants and animals live together as a harmonious whole, joyfully and in mutual benefit.

As explained earlier, in the desert there are many places with rivers and underground water. One method for beginning a natural farming project in the desert is to revegetate the banks of the rivers and then gradually work outward from there to make the interior areas green. If we establish trees and other vegetation along the rivers, their range will naturally expand. If possible, however, we should scatter every kind of seed over the entire area at the same time and revegetate the desert all at once.

The theoretical basis of revegetation from the riverbanks follows the "plant-based irrigation method." It does not rely on running the river water through concrete waterways, but encourages the water to follow greenbelts of plants. It achieves non-irrigation agriculture by increasing water retention in the soil and the plants themselves.

Water naturally moves to lower areas, is carried by the roots of plants, and creeps toward dry areas. At the river's edge, reeds and cattails will flourish, while species of arundo grass (*Arundo donax*) will grow in clumps, protecting the banks. Pussy willows, purple willows, and alders will provide protection from the wind, cool the understory, and draw water.

If we plant every kind of plant, starting from the area around the river, the underground water will filter up the roots of the plants, and gradually a protective forest will develop. This is what I call plant-based irrigation.

For example, if you plant acacia trees sixty feet apart, in five or six years the trees will reach a height of thirty feet and the roots will have spread at least thirty feet in every direction, carrying the water with them. As the soil fertility increases and humus accumulates, the soil's ability to retain water will increase. Although the movement of underground water is slow, gradually it will move from one tree to the next, and they will act as water bearers.

If we were to apply this method for revegetating the desert, we would begin by planting woods along the rivers in the desert. Then, at right angles to the river, we would create greenbelts of natural forest instead of irrigation canals, and have them serve the role of waterways. In the center of these greenbelts, we would plant fruit trees and vegetables, creating natural farms. In this way, we would be creating food and rehabilitating the desert ecosystem at the same time.

You may think it reckless for me to say that we can revegetate the desert. Although I have confirmed this theory in my own mind and in my orchard and fields in Japan, I have had few opportunities to prove it on a large scale. Recently, however, the government of India asked me for technical assistance in carrying out an aerial seeding effort. I told them, in general, it is best to leave nature to nature, and let nature recover on its own. Sometimes, however, as in this case, if the land has been too badly damaged, we must provide nature with the materials it needs to become healthy again. I agreed to help with their efforts.

In India, there are more than five hundred varieties of trees that bear edible nuts and more than five hundred varieties of fruit trees. In addition to these, we should sow the seeds of five hundred varieties of grains, vegetables, and green manure, ideally on the barren Deccan Plateau and in the desert. No matter how bad the conditions are, some varieties will be suited to that place and will germinate, even if some die. These "pilot" plants will help create conditions that will allow other plants to follow.

A second purpose in sowing a variety of plants and microorganisms is to awaken the sleeping earth. There are some deserts, particularly the sandy ones that have apparently lost their ability to support life and have all but died. Many savannas, however, exist in relatively young, clay soil. These deserts contain all the nutrients plants need, but for various reasons these nutrients are unavailable to them. In order to make these nutrients available, to rouse the earth, a variety of ground cover plants and microorganisms are necessary.

The earth will not come back to life if we only plant a small variety of trees we deem to be useful. A tree cannot grow up in isolation. We need to grow tall trees, midsized trees, shrubs, and understory plants all together. Once a mixed ecosystem is re-created, the rain will begin to fall again.

It seems logical for people to choose something special from nature and use it for the benefit of human beings, but when they do this, they make a big mistake. Taking one element from nature, in the name of creating something valuable economically (cash crops, for example), gives that element special value. It also implies that other elements have a lesser value. When human beings plant only "useful" trees with high cash value in the desert, and cut down the undergrowth referring to those plants as "weeds," many plant species are lost. Often they are the very plants that are enriching and holding the soil together.

There is no good or bad among the life-forms on earth. Each has its role, is necessary, and has equal value. This idea may seem simplistic and unscientific, but it is the basis for my plan to regenerate landscapes all over the world.

The Revegetation of India

Although I went to Africa, India, and the United States sharing my ideas, and experimenting on several large areas, I did not have the means to accomplish what I had set out to do. Back home from my travels, I was bemoaning the futility of my efforts when, on Christmas night, a woman named Kimiko Kubo, of Chiba Prefecture, came to my hillside hut. "As thanks for the elephant Hanako that Prime Minister Jawaharlal Nehru sent to the children of Japan after the war," she said, "I would like to give a small gift to the children of India." After that, she put a large sum of money, wrapped in newspaper, into my rice bin and left.

I immediately thought of how I could put this money to best use. I decided to use it to help carry out my dream of regreening the deserts of India. First I flew to Thailand to collect seeds. When I had traveled to Thailand a year earlier, I had visited a primeval forest, and I thought it would be a good place to collect seeds for the deserts of India. A Buddhist monk I met there had promised to obtain official cooperation for this seed gathering, and I had also received an informal invitation to meet with the royal family during my next visit to the country. If I was able to get permission to collect seeds in the ancient forest, which belongs to the royal family, then I could have some young monks gather them for me and be on my way . . . or so I thought.

When I got to Thailand, the world was in turmoil, and talks went nowhere. As we were planting seeds on the grounds of the Kanchanapuri Children's School, the Persian Gulf War broke out, and plans for starting a seed bank in Thailand vanished. I returned to Japan after discussions were suspended. During the six months that the world was in an uproar, I put together reports for use in the deserts of India and waited for an opportunity to take them there.

I also wrote a few articles saying that rather than bombs, it would be better to sow seeds in clay pellets from airplanes that had previously been used as military bombers. When the *Asahi Shimbun*, a respected national newspaper, and the *Ehime Shimbun*, a local newspaper here in Shikoku, carried the story, people in Japan became interested in sending seeds to the people in the deserts of India, Africa, and other places. Housewives and children from around the country began sending seeds from fruits and vegetables they had eaten. Also, with the cooperation of the Murata and Sakata seed companies in Matsuyama, and the help of local environmentalist Masao Masuda, the gathering of seeds progressed nicely. When the seeds and other donations had been collected, I felt that I could at last return to India.

I called upon the help of the Tagore Society,* which had made arrangements for me before. Accompanied by a woman named Sister Nagashima,

---

* The Tagore Society is a worldwide network of local groups dedicated to carrying out the philosophy of Rabindranath Tagore, the great humanist, writer and poet, philosopher and Nobel laureate. The mission of the group is to fulfill Tagore's dream of a "borderless world."

who had previously interpreted for me, I set off, leaving the rest up to chance.

First, I called on the governor of the state of West Bengal and viewed the site where we did an aerial seeding of mangroves at the mouth of the Ganges River during my last visit. We went up the vast Ganges for about one hour in a sailboat from the Ministry of Forests and Environment. After transferring to a motorboat, we plied the shallow waters to our destination. The seedlings of twenty to thirty varieties of mangroves had sprouted on the sandbar and grew as far as the eye could see. I shared my amazement and delight with my guide, Mr. Dasgupta.

When I pulled one of these plants up, a number of small shellfish and hermit crabs crawled out. I was amazed by these tiny creatures that made their home in this great river. I wondered why this example of success had not spread to the other countries of the world concerned about the disappearance of the world's tropical rain forests. Whether it is because India is an isolated country or because the academics have not yet heard about it, I do not know. I *do* know that this example shows the usefulness of aerial seeding for revegetating large areas in a short period of time.

The office of the officials working at the site was on the riverbank. As we neared it, the boat became mired in a muddy spot, and we could not get out. At that point some local fishermen gathered and pushed the boat, with us six passengers still aboard, up onto the bank. I stopped briefly inside the building, then went out to the garden and saw a beautiful multicolored image of a Hindu goddess. The laborers proudly told me she was their guardian deity of the forest. These workers lived in houses with about 150 square feet of floor space.

What interested me most was that there was wild rice growing in a marshy belt along the riverbank. This rice had grown up on its own without being planted. The soil was fertile, and a rare alga that produces nitrogen and potassium was growing over the entire surface. These were considered wildlands and were not being used. The rice seemed to be a primitive form, with forty to fifty grains per head.

If we were to sow pellets of unhulled rice directly from airplanes in areas like this, success would be almost certain. I felt that there were many productive things we could do right away there in Bengal alone. I wrote an impromptu poem with a brush on a large sheet of paper, to hang in the cabin of the boat. I was pleased that well-formed characters emerged so smoothly from the tip of the brush.

When I returned to Calcutta, I heard that many people were waiting to see me in Bangalore, in the south, but I could not go because of previous commitments, and sent Mr. Dasgupta and Mr. Makino in my place. Instead, I flew to New Delhi with my secretary, Ashoka, and Sister Nagashima. Mr. Singh, of the Office of Forestry, introduced me to an official of the Ministry of the Environment. Within the Ministry of the Environment there was an Office of Wasteland Development, and a Forest Division within that. I spent a day discussing my plan for the revegetation of the Indian desert with the official, who was a very positive, active person, and was sincerely interested in the health of the environment. He asked me if I would like to meet the prime minister, but the premier of China had just arrived in India, and I thought a meeting would be impossible in the midst of such important discussions. Even so, it was arranged for us to meet for fifteen minutes the next day.

Because I would only be able to spend a short time with Prime Minister Rao, I took a small container with me containing rice I had grown on my farm, and seeds of Egyptian clover suited to the tropics, which were a gift from the Sakata Seed Company. I began by asking the prime minister to promote an agricultural revolution and to encourage the revegetation of the desert through the spread of the clover. This caught his interest, and he occasionally picked up the container or the rice and looked at it carefully. We talked for almost an hour. I later heard that he had studied agronomy when he was younger and was well versed in agricultural methods. The prime minister then asked that I be introduced to the head of the Ministry of Agriculture.

The following morning, I talked with technical experts, perhaps as a sort of trial, and in the afternoon I met with the minister. Our discussion centered on the differences between scientific agricultural methods and natural farming.

Fortunately, during this unexpected week in New Delhi I was able—thanks to the excellent interpreting of Sister Nagashima—to enjoy refining my ideas with a number of government officials. Accounts of these events were carried throughout the country on national television and in the newspapers. The prime minister's support of natural farming was also widely reported in the media. I felt that great interest in the revegetation of India had suddenly arisen.*

From New Delhi, we went to see a site at Gwalior, in the state of Madhya Pradesh, where aerial seeding had been carried out that summer. Then we flew to Agra in a former military airplane and were taken from there to the Office of Forestry. After riding in a jeep across a plain dotted with fields of rapeseed plants and sugarcane, we arrived at the Chambal Gorge. There, the scenery underwent an abrupt change. Before us was an odd desert as far as one could see. It was bare, red earth hills ranging from several feet to about a hundred feet tall.

There were scattered stands of acacias and other trees in the bottomlands with only a few grasses and a little sheep sorrel (red sorrel, *Rumex acetosella*). The barren clay earth in this desert made revegetation seem a daunting task. Even if seeds were sown, germination of naked seed would be poor. I was told that when rain fell, the seeds would probably be washed away. Therefore, the soil had been deeply tilled and a number of trenches, or swales, were created. I could only imagine how much effort those trenches required to dig by hand. Despite those efforts, there were hardly any plants growing where the seeds apparently had been sown earlier.

I was told that the seedlings had been eaten by goats and other animals. One of the workers laughed ruefully and said, half in despair, "Well, even if we don't get a forest, the livestock are getting fed."

---

* Even today, interest in Mr. Fukuoka and his natural farming techniques and philosophy is probably greater in India than anywhere else in the world.

Just then, as the jeep rounded a corner on the rutted road, a herd of goats came pouring out in front of us. The herdsman, startled at the sight of the gun carried by our guard, desperately flailed his whip and, as if fleeing, chased the goats up a bank that seemed impossibly steep. The sight suggested to me that people were not allowed to move about freely in that area. The farmers, who live with cows and goats, seemed to have sneaked into the area at times in search of the scant vegetation, and the officials could not avoid giving tacit approval.

In any case, I imagined that revegetating this area would be very difficult, but I could see that at least some of the seeds that had been sown previously in clay pellets were growing well, and for that, I was relieved. Afterward, I heard that in the Madras area clay pellets had been sown successfully, and I made an agreement with our pilot to go there a few days later.

At first I thought that if four or five years of work had only resulted in this much vegetation, then creating a forest where elephants could live—which was the long-range objective—was a long way off. Still, as I studied the plants in the area and listened to the exhaustive explanations of the government officials on the site, I realized that, even though the results were not as obvious as I had expected, we were doing the right thing. Sowing a variety of seeds in clay pellets *would* work here.

It would have been easy for me to talk about the conclusions I drew at the Chambal Gorge then and there, but I simply praised the results achieved through the strenuous efforts of the people of the state government without giving any criticism, and kept my thoughts to myself. This difficult site needed a bit more thought.

From the time we arrived at our next destination by jeep, a number of children and old farmers gathered around us and intently followed our every move. When I looked into their eyes, they seemed to want to tell us something. I had the feeling that they knew the true cause of desertification. Their eyes were the same as the eyes of the children I met in Somalia, who, when given a handful of seeds, immediately set to work and produced excellent vegetable gardens. I felt that the children of India could do the same thing.

I was told that ten years earlier, there had been elephants in this area; just three years before, a tiger came into the village. But as a result of desertification, the villages downstream had disappeared, the vegetation had disappeared, and the elephants and tigers no longer came. The desert had advanced at an unbelievable speed. Before working out measures against desertification, it is necessary to ask why the area has become a desert—to search out the true cause, and to cut that cause off at the root. We must begin by understanding that it is people's actions that have caused the expansion of the deserts.

Although the results were not as I had expected, as we went around by jeep, I kept catching glimpses of small green areas in the shadows of large rocks with clumps of grasses. I wondered: How did this vegetation survive in these harsh, dry highlands?

As we were returning, the sun was beginning to set, but someone suggested that we look at the crocodiles in the river. So we boarded a motorboat and headed out. The river was more than three hundred feet wide, and the water was remarkably clear. We rode for some time, but there were no crocodiles. Indeed, they had known from the start that there were no crocodiles there and simply wanted an excuse to enjoy a boat ride. No matter, I gazed at the desert scenery in the evening light. Many things crossed my mind that evening about the vast drama of human history.

When I returned to the Office of Forestry the next day, I visited their garden and was allowed to hold a small crocodile and a number of other rare animals that had been bred in captivity there. I was told that there were now only fifty Indian crocodiles left in the Ganges. I realized then there was a profound connection between crocodiles and the deserts of India.

If there are rivers, why are there deserts? What was the key to solving this puzzle? On my boat ride I saw that the water in the river was clear, yet there were no crocodiles or fish. In Somalia, the Juba River flows through the desert, full of water year-round, and catfish live in the muddy water. The problem, as I have said, is not that a place becomes a desert because there is no water; nor is it the case that if there is water there will inevitably be fish. The relationship of soil, water, trees, and human communities is not as straightforward as specialists would have us believe.

Even in a jungle where large trees grow and elephants exist, once the ax of human beings is introduced, the trees disappear, and the land becomes a grassy plain. When people begin to live there communally, raising goats and cows beyond the land's capacity to support them, the green of the plain fades almost instantly. When rain falls on the bare soil, mudslides occur and the fertile soil is washed away. All that is left is a rugged wasteland. Without vegetation, the earth dries out and becomes a desert.

In the attempt to restore the vegetation in a desert created in this manner, scientists would have us first pour water into the desert to bring the vegetation back to life. I suggest that the result of such action will result in further damage.

In order to explain why this is so, I first must discuss the relationship of water and soil one step farther. Originally, water, soil, and crops were a single unit, but since the time people came to distinguish soil from water, and to separate soil from crops, the links among the three were broken. They became isolated, and were placed in opposition to one another.

Water in which organisms no longer live is no longer real water. Soil without grass, although it is called soil, is not really soil. Earth without grass loses its connection to water and becomes parched. It is not surprising that grass and trees will not thrive there. So, instead of thinking that grasses and trees grow in the soil, it is actually the grasses and trees, other plants, animals, microorganisms, and water that *create* the soil and give it life.

When Western scientists see a dry desert, they introduce irrigation. To accomplish this, they interrupt the natural flow of water. They build dams, collect water, and build canals and waterways to transport the water. The most commonly used method to get the water into the irrigation channels is to raise river water by using pumps. When I visited Somalia, I saw that the Russians, the English, and the Italians in the past, and modern farms supported by France and Japan, all used this method. They carried it out as follows:

Soil from the area around the river was piled at right angles to the river using bulldozers to construct dikes more than thirty feet high. The longer the distance, the higher the dikes needed to be. On top of these, dirt or concrete waterways were made to carry the water pumped up from the river. Small fields below the dike were watered in succession. Water is precious, of course, so only a small amount is released into the fields at one time. The water soon evaporates, and all that is left is salt. After just a few years, the site is abandoned and the farm is moved to a new place.

In return for a minimal amount of food, the refugees and other people who settle on the edges of these large farms spend their days working in pairs, using a scoop attached to a rope to remove the sand blown into the fields' shallow ditches. This futile labor is the sad reality of desert life.

No matter how diligently people try to grow crops in the barren earth from which the topsoil has been scraped away, they cannot grow enough make a living. Plus, the monoculture cash crops they are made to grow are weak and susceptible to insects and disease. When the farms change location, not a single tree or blade of grass remains. All that is left is a flat desert, worse than before, and a big pump that is out of gas.

I cannot imagine why they build waterways on top of these strange, high dikes in the first place. But when you think about it, if the water is raised to such heights, and the authorities are the only ones who control the pumps and water rights, then the people of Africa are having their lifeline of water wrested from them. Such waterways proved to be an effective means of control by colonial rulers. The same strategy is being used today by agribusiness to take farming away from the farmers all over the world.

A more appropriate method of irrigation would use the natural flow of water as much as possible. Large rivers have tributaries that flow downhill, and the tributaries are fed by smaller streams. If ditches are made from these small streams, water can be run directly into the fields. Solutions like this are simple, effective, and appropriate. They allow us to live peacefully within nature without going to a lot of trouble.

After returning to New Delhi, we headed for Calcutta, flying at night in a dense fog. In the cabin we heard an announcement that due to the fog, we were flying by a different route, and we landed in Hyderabad, in the south. If we could have left the plane, we could have seen the results of the aerial seeding nearby, but we were under strict orders to remain inside. Then we went on to Bombay, ate breakfast on the plane, and took off again, finally landing in Calcutta. As a result, I had an entire day of free air travel in the skies over India.

I discovered that there was almost no vegetation between Bombay and Calcutta. I later flew over Bangladesh and Burma. To my dismay, I could not see anything from the plane window except desert until we approached Thailand. If people were to see detailed aerial photographs of the earth, they would understand just how serious the current environmental problem really is.

From Calcutta I was able to go to the state of Manipur, thanks to Professor Makino, who teaches at Manipur University. Manipur is almost completely closed to foreigners.* During World War II, Japanese soldiers entered the capital city of Imphal from Burma and were all killed in a battle with Somali and Ethiopian soldiers fighting under the British flag. When I stood before a marker raised by the local people to commemorate the battle and heard an account of how the Japanese soldiers fought, I was overcome by the cruelty of history. In a commemorative hall near the floating island, there was an exhibit of many photographs of Japanese soldiers working in cooperation with Chandra Bose,ᵏ a champion of Indian independence. I viewed the exhibit with mixed emotions.

In this area, the people and climate are similar to Japan's. The former king, and present governor, looked just like the feudal lord of Matsuyama near my home. He welcomed us and gave us an opportunity to speak before other officials and university students. Thanks to his good offices we were able to freely tour the state without any difficulties.

One day it rained, so we took shelter in an elementary school nearby. Some children gathered around us, and it occurred to me to show them photos of my natural farm. As I did so, I told them about my

---

* Manipur, or "land of jewels," is a small state in northeast India bordering Burma (Myanmar). Because of its remote location and political unrest from groups hoping to reestablish Manipur as a sovereign nation, travel had been restricted only to those who received a Restricted Area Permit. That restriction was removed beginning January 1, 2011, for a period of one year.

ᵏ Chandra Bose was one of the most prominent leaders of the Indian independence movement. He and his Indian National Army fought with the Japanese against the British in the Battle of Imphal in 1944.

natural farming method. I told them that it should be possible to sow seeds of tropical fruit on the barren hills in the area, and to grow many of the thirty kinds of fruits that I grow in Japan. I encouraged them to create a paradise such as the one they saw in the pictures. The soil in their fields is fertile. If they were to directly sow a variety of seeds in clay pellets, they would not be troubled by the deep mud. When I told the children to go home and explain this to their mothers and fathers, they listened with shining eyes.

The governor of the state also joined in and said that I was telling them the truth, which caused quite a stir. They were already overwhelmed that the former king was there in the first place. Seeing those barefoot children gathering firewood in the rain reminded me of my own elementary school days. Their eyes sparkled like those of the children in Somalia and at the children's school in Thailand.

On my final day there, a roundtable discussion was held in a public hall. At the end of my talk, I read a poem I had written about the lovely floating islands of Lake Imphal. I said that if I were asked to choose one photograph that I had taken in my travels around the world, I would choose a photo of those islands. There was a burst of applause, and one of those present said, "We had thought that we were the poorest, most insignificant country, but you have told us that this country could become an ideal home for the world. That has given us courage." Then they wrote a declaration of their intention to create a paradise there and had me sign it, too. I had a very tight schedule during my time there, but every evening I was invited to a banquet and entertained with lovely songs and folk dances. It was certainly an enjoyable, worthwhile visit.

It was nothing unusual for the plane to Calcutta to be grounded for days at a time, so we went to the airport early in the morning to make sure there even *was* a plane. It was indeed there but had a mechanical problem that could not be easily repaired, so it was arranged that another plane would come for us in the afternoon. We had no idea when it would arrive, and the information we received raised our spirits and then lowered them several times during the day. The rainy airport was cold, and even though the three of us covered our heads with coats, we were still shivering. Nevertheless, ten brave souls from the Ministry of Agriculture waited to the very end to see us off.

Finally, our plane arrived in Calcutta, and we set out for Thailand again. I stopped there on my way home to check on the status of our efforts to create the seed bank to guarantee a large supply of seeds to carry out our aerial seeding plan on a large scale. But things were unchanged in Thailand, with political negotiations going one step forward and then falling back. All I could do was lament my limited influence. But steps forward were made in India, and the revegetation of the desert is closer than before.

Notes from an International Environmental Summit

Close to the time of this writing (1992), an environmental summit was held in Brazil, and a number of environmental problems came under discussion. I could not very well let this go by without somehow participating, so I attended the following meetings, which were held in Kyoto ahead of time.

1. In June, the day after the meeting of intellectuals organized by Mr. Takeshita for Maurice Strong, the director of the Brazil summit, a panel discussion, moderated by Ryu Tachibana, was held at the Shiba Zojoji Temple. The discussion was broadcast nationwide on NHK.*

2. A one-day meeting at the Teikoku Hotel in Kyoto with the minister of the environment of India, to refine plans for a movement to create a forest for elephants.

3. A meeting with Mr. Runphal of the Brazil summit, the three major newspapers, and their editorial writers. This meeting was sponsored by the Kyoto Forum.

4. A meeting with representatives of the Magsaysay Foundation of the Philippines.

When I met with Mr. Strong, I showed him a head of rice I had grown in my fields and told him that even if the population of the world doubled, we could still feed everyone through natural farming, without using a drop of oil. With Mr. Runphal, I only talked about concrete steps for desert revegetation.

---

* The national broadcasting company of Japan, like the BBC in England.

Then I met with the Indian minister of the environment to discuss the conditions that would be necessary to carry out the revegetation of the deserts. Then I met Mr. Singh, along with people from the embassy, to work on plans for creating forests for elephants.

Here is what I suggested:

First, that an organization be created to ensure that seeds and funds for use in revegetation collected from the Japanese people be directly and securely put into the hands of the farmers in India who will sow them. To do this, the seeds should be exempt from quarantines, or at least the procedures should be simplified.

Second, the Indian government should open the large tracts of "wasteland" they control, without charge, to people who will sow seeds on them, or, as in Japan, lease the land for a nominal charge.

The reason Japan was successful in replanting forests throughout the country is that seedlings were given to farmers free of charge, and the forestry cooperative was always there to back them up. In India, seeds should be distributed without charge, and the seed bank should always have seeds on hand.

Application procedures for sowing seeds in the desert should be simple. In Japan, farmers simply indicate on a single sheet of paper the date, place, and area where they want to plant trees and turn it in to the forestry cooperative. Afterward, the cooperative confirms that the work has been completed.

The way it is now with the assistance program for rehabilitating wasteland in India, the farmers must submit a detailed plan twenty-five pages long. You have to record, in minute detail, who is the head of the business, what class they are from, who will receive the profit, if any, how responsibility will be taken in case of failure, and on and on. No farmer, on top of all the work already on his or her plate, would realistically be able to do anything under such a system. I asked that more trust be given the farmers and procedures be made simpler, but that point has not yet been settled.

CHAPTER SIX

Travels on the West Coast of the United States

THE FIRST TIME I VISITED the United States, it was at the invitation of members of the natural food movement. The second time, seven years later, I was invited by the state universities of Oregon, Washington, and California, and by the leaders of the international permaculture movement. I was to be a guest speaker at several conferences on natural farming, organic farming, and permaculture. It was a very busy schedule for both visits of about six weeks each time.

On the second visit in 1986, I set out with the hope that in the United States I might be able to come to a better understanding of the problems I had not yet solved in Africa. The people directing the conferences had arranged my schedule with great care. They had done thorough groundwork and had scheduled every free moment in order to show me things they thought I would be interested in and might be useful for me. They also wanted to learn as much as they could from my experience.

The plan for this second trip was for me to start in Washington State, travel around Oregon, go on to California, and then fly to the northeastern states of New York and Massachusetts. Priority was given to the international conferences and to lectures I would give at other universities and to other groups. Once or twice a day I would visit a farm—usually an organic or natural farm, but not always—meet the farmers, and hear about their operation. If I were asked, I would give my impressions and sometimes practical advice. The travel time by car would often be two or three hours each day, and on several occasions we traveled in a small airplane. Between events, I gave interviews for newspapers, radio, and television, so I had very little free time. Afterward I thought it was quite a feat that I kept this up for almost fifty days.

Ignorance is bliss, and since I am an easygoing fellow, I just played it by ear. When I stood up to talk, I would look at the expressions on the faces of my listeners and then say whatever came into my mind. I could talk for hours without getting tired.

As I traveled from north to south on the West Coast, I could see that desertification was advancing with increasing speed. To the north, in Washington, there are still some old-growth forests, but as you enter Oregon, the trees on the mountainsides gradually become more sparse. The grass disappears and the green of the landscape grows fainter. This is closely connected with the deterioration of the soil and poor management.*

This time, as I traveled about the United States, my impression was quite different from that of my visit seven years earlier. It seemed that the environment surrounding the agricultural areas had deteriorated since I had been there before. I often commented on how the green that was there was an imitation green, how the agricultural products were simply petroleum by-products, and how the future of American agriculture appeared to be in danger. This time I felt even more strongly that my foreboding was correct.

About half of the land in the United States is, or is becoming, desert. I felt that the expanding American desert was at least as great a problem as the deserts of Africa, but most Americans seemed totally unaware that their country is becoming more arid. As I discussed previously, they think it is only natural that when little rain falls in the summer, the grass dries up and the plains turn brown, but it was not always this way. Americans are so dazzled by the vastness of their land that most people do not seem to be concerned about preserving it.

I have heard that these days many American farmers, large and small, are in dire straits. The farmers seem convinced that the reason for this is the decline in

---

* The West Coast of the United States has a Mediterranean climate, which means that rain falls in the winter, but not during the summer. This effect is most extreme in California and decreases as one travels north. The forests of the Maritime Northwest in Oregon and Washington are cooler and more humid, so they are better able to store nutrients in the soil and in the litter layer on the surface of the soil. Damage to soil in a drier climate, like California's, is more difficult to repair.

agricultural exports, but there is no way, amid the ruin of the land, that farmers can become well-off no matter how much petroleum "rain" they use to grow their crops. Unfortunately, this point is not yet being given serious consideration.

Agricultural experts and agribusiness are bound by the idea that even land that has lost its natural vitality can still produce crops with the addition of petroleum energy, agricultural chemicals, and water. On this point, Japanese scientists and farmers have the same belief, considering this form of agriculture to be "advanced." Some actually consider the soil to be a nuisance.

People told me that in the United States, reform and innovation always seem to begin on the West Coast. This time, when I visited the three westernmost states of California, Oregon, and Washington, I could see and feel the beginning of an agricultural revolution among the farmers of this region.

Farmers' Markets

It appears that in America today a revolution in consciousness is occurring, not only in agriculture but in every area of life. I could feel it both in the cities and in the farming towns, where I was received as an agriculturalist as well as an Asian philosopher.

I especially enjoyed the weekend farmers' markets. In some green open place, like a park or college campus, people were selling gift items and toys along with an impressive variety of locally grown produce amid colorful banners, street performers, and lively music. The booths were beautifully decorated, and cheerful, energetic youths called out to customers in animated voices. Everyone seemed to enjoy meeting with their neighbors in the market's relaxed, exhilarating atmosphere.*

There were booths serving ethnic food, such as Italian, Indian, and French. Caucasians dressed in Japanese *happi* coats^k enthusiastically rolled sushi and sold Japanese pickles and fresh tofu. Some of these shops offered delicious tidbits of tofu prepared in ways we do not often see in Japan. Everything was unique because it was improvised and made by creative food enthusiasts.

Of course, the heart of the market was made up of all sorts of fruit and vegetable stalls, as well as shops selling things like dried fish and pancakes. I was surprised that all of the food sold in the market was organically grown. Products that can commonly be found in supermarkets were excluded. This is at the request of the consumers.

Because the farmers' market was maintained as a place for natural farmers and local merchants, the salespeople made every effort to tell the customers all about the varieties of the grains and vegetables and the way they were grown. Profit was not the main motivating factor, so ingenious, imaginative products were sold with a whimsical spirit.

---

* Comments in this chapter about things like the atmosphere at the farmers' market, the look of the towns and cities in the United States, and the demeanor and diet of Americans as opposed to Japanese, are meant primarily for Japanese readers who have never been to the United States.

^k A traditional Japanese coat, often with a decorative crest, which is usually worn to festive events.

Each item was different, and, unlike those sold in a supermarket, they were all fresh. People would spend an entire day at the market, enjoying shopping, eating lunch, listening to music, and napping on the grass. Parents brought their children and made it an all-day outing. Then, before leaving, they would buy a week's supply of fruits and vegetables and take them home. Understandably, the popularity of these markets is skyrocketing.

When I walked around looking at things in one of these markets, people frequently called out, "Hey, aren't you Fukuoka?" They looked curiously at my traditional Japanese work clothes and presented me with unusual apples and pumpkins. At the markets in places like Davis, California, and Eugene, Oregon, it seemed that every third person knew who I was, and soon the arms of the people accompanying me were filled with presents. After a while I would get caught up in the mood and would draw one of my clumsy philosophical sketches with a felt pen or an ink brush on a piece of cardboard or whatever was available. When I gave these to people, they were delighted and promptly displayed the signs at their booths. These philosophical drawings were also used as posters at the entrances of the university auditoriums where I spoke and outside natural food stores. I was amazed to see these images even being given away on T-shirts.

The atmosphere at these farmers' markets seemed decidedly Asian to me. I think Americans feel isolated somehow by their individualism and enjoy the congenial, openhearted atmosphere usually associated with Eastern cultures.

Another thing I could not overlook is how rapidly the natural foods movement is spreading across the United States. This is largely due to the efforts of Herman and Cornelia Aihara, in Oroville, California, who first invited me to the United States in 1979, and also to Michio Kushi, in Boston. When I came for this second visit, I could see that country-style Japanese food and other Asian cuisines had taken hold throughout the country as the standard for a delicious and healthy diet.

The flavors and ingredients of the sushi and tempura shops in the United States were very good, perhaps even better than in Japan. People are completely at ease with the Asian way of eating the food, and are beginning to comment on the choice of sushi ingredients, the way it is made, the tempura ingredients, and the quality of sauce for dipping noodles. Today, even a typical home kitchen might offer miso soup with brown rice and fresh steamed vegetables prepared in an American manner.

Given this situation, farmers are naturally going to set their sights on providing ingredients in response to people's preferences. When people become skilled at cooking, they seek a variety of ingredients. Amid this natural foods boom, produce grown by organic and natural methods is in demand. Americans seem to be rapidly turning their attention more toward delicate, subtle flavors, and a more natural lifestyle. And they are planting many more vegetable gardens at home.

You might say this is a revolution occurring from the ground up.

Urban Natural Farms

Residential lots in the United States are generally quite large, especially when compared with those in Japan. The average house may have a natural growth of trees and a lawn, so when it comes to creating a family vegetable garden, unlike in Japan, it is possible to have just the kind you want.

On the ideal natural farm or urban homestead, there would be a mixture of fruit and nut trees, and beneath them vegetables, grains, and berries. Chickens would run around in the weeds and clover. When I talked about such things in Japan, I was considered unrealistic, but in the United States this idea is easier to understand for most people, and easier to carry out. When I suggested that it would be a good idea to plant fruit trees to line the streets in towns and cities and to grow vegetables instead of lawns and annual flowers, so that when the townspeople were taking a walk, they could pick and eat the fruit from the roadside, people were surprisingly enthusiastic.

When I suggested that it would be good to scatter the seeds of clover and daikon on the existing lawns so that in two or three years the clover would overcome the lawn and the daikon would take root amid the ground cover, interestingly, it was the Asian people and Asian Americans who said they would try it right away.

Most Americans would just laugh and agree with the theory, but they were cautious about putting it into practice. The reason, I believe, is that it would challenge their adherence to "lawn" culture. If they cannot overcome this prejudice, there will be a limit to the growth of family gardens in the United States.

It seems that the main goal in the life of the average American is to save money, live in the country in a big house surrounded by large trees, and enjoy a carefully manicured lawn. It would be a further source of pride to raise a few horses. Everywhere I went I preached the abolition of lawn culture, saying that it was an imitation green created for human beings at the expense of nature and was nothing more than a remnant of the arrogant aristocratic culture of Europe.

People Sow and Birds Sow

Breitenbush Hot Springs was the meeting place for a three-day symposium about permaculture and natural farming. The conference center was beautifully situated in a forest in the foothills of the Cascade Mountains of Oregon. There was a large meeting hall, a dining hall, and a number of small cabins scattered among the trees. We had a relaxing time soaking in the hot springs and enjoying the geothermal sauna the night before the meetings began.

In one corner of the property, a natural vegetable garden had been created to provide food for the dining hall. The person who began the garden, a young man named Katsu, had left a few years earlier, and the garden had been neglected. When I heard this, I decided to hold a training session there.

Katsu happened to be accompanying me on this trip as an interpreter. This was the first time he had returned to the garden since he had left and let nature take its course. According to Katsu, he had followed the natural farming method, planting clover over the entire area, dividing the many vegetables into small plots, and leaving them alone after they were planted. Seeing the garden now with its riotous growth of clover, weeds, and vegetables, he thought the experiment had resulted in failure.

At first glance the field appeared to be a failure to everyone, with the plants growing in such confusion. But when we looked more closely, clover had spread over the entire garden, there were not as many weeds as there appeared from a distance, and in the midst of all this, vegetables were growing hidden in the overgrowth.

I told them that this field is actually doing just fine for the first stage of a family garden based on the natural farming method. In the first year people sow the seeds, in the second year nature makes adjustments, and in the third year many unplanned and unexpected surprises begin to appear. That is when nature begins to make a natural garden for us. Thanks to Katsu's neglect, the vegetable flowers had bloomed, gone to seed, and reappeared as volunteers. Other seeds had been eaten and dispersed by birds and mice, and, as a result, the vegetables were scattered in every direction.

Within the garden, however, there were dry areas, damp areas, shady areas, and areas where the soil was more fertile than others. For this reason, not all of the seeds that fell to the ground succeeded in growing. Only the seeds that landed in appropriate places sprouted at the appropriate time the following year. The seeds that were meant to die, died, and only those meant to live grew up and thrived. Sure, the garden looked like a jumble, but gradually nature would show which plants would grow well in which places. The clover was spreading nicely, and enriching the soil, and it would gradually help keep the weeds from getting out of control. If an effort was made to control a few special weeds by hand, in time a fine field would emerge.

We continued admiring the garden Katsu had created. Here and there stalks of wheat and rye appeared that were not intentionally planted. The number of seeds per head was greater than one might have expected. This suggests that it would be worthwhile to scatter seeds of rye and other grains in this field as well. Crops like garlic grew well there, so we ought to be able to get a succession of rye in the spring, followed by lettuce, tomatoes, melons, and cucumbers, then leeks, and in winter, broad beans. All of these would be growing in a permanent white clover ground cover.

In any case, this garden, which appeared at first to be a failure, was actually succeeding quite nicely. When I gave this explanation to the group, everyone was delighted. But the one who gained the greatest confidence was Katsu.

Like Katsu, many people in the United States are equally practical in their approach to things and are quite down-to-earth. The American people have an openness about them that comes from enjoying the natural world. I met many who were broad-minded enough not to be perturbed by a few weeds or disorderly fields that, in my experience, would have been intolerable to a traditional Japanese person.

All in all, it seems that natural farming has made a solid beginning in the family gardens and small farms of America. Because residential lots are large in the

United States, a family vegetable garden can provide for all the food needs of a typical family, if they are willing to do the work.

In Japan, a residential lot of about a quarter acre would be enough to allow near self-sufficiency and provide a healthy living environment, but I learned—to my envy—that in many suburban and rural areas of the United States, people are not allowed to build houses on small lots. While each state and country are different, in one part of upstate New York where I visited, they are not permitted to build on *less* than one and a quarter acres.

Where residential lots are large, the living environment changes completely. If the Japanese were to go into the mountains, enjoy the freedom of nature, and strengthen their spirit of independence as Westerners do, real estate prices in the cities would plunge.

In the United States, we sailed along on the highways through brown plains that are turning into deserts, but when we entered the towns, we suddenly saw a luxuriant growth of shrubs and large trees. The typical town we visited was quiet, clean, and full of greenery. Wherever you go, the middle-class towns and suburbs are so covered with vegetation, you can barely catch glimpses of the buildings. In many cases, the homes are not visible at all. Wealthy people seem to want to live far into the hills. The finest houses are often situated deep in woods or forests. In the large cities, where many poor people live, there is very little greenery. This is just the opposite of Japan where the wealthiest people live in the cities. The difference,

I think, is that the Japanese do not have the sense of freedom of the people of the West.

While I was riding along the highway from Oregon to California, I wondered aloud what would happen if we were to broadcast seeds of daikon and clover from the roof of the car onto the brown wasteland on either side of the highway. A young man riding in the car with us took packets of various kinds of seeds out of a bag he had with him and presented them to me. He said he was a "barefoot"* botanist who specialized in mycorrhizal bacteria and nitrogen-fixing plants. He had heard me talk several days earlier and had come along because he was impressed with the campaign to sow seeds in the desert and wanted to help.

Just as we entered California, we stopped the car in a turnout on a mountain pass overlooking a broad dry landscape. He handed me the seeds and asked me to sow them. I broadcast the seeds from the top of the pass into the valley below. The others did the same while crying out in delight. The seeds were lifted by the wind and carried far away.

He was not the only one. Throughout the Pacific Northwest, I met many groups and individuals carrying out research that is generally considered unimportant to the general public. They are collecting the seeds of heirloom varieties of vegetables and other food-producing plants, along with others that are important to the ecology of the region even if they have no apparent commercial value. They are propagating these

---

* This expression refers to someone who is self-taught as opposed to someone who has academic training.

plants, collecting the seeds, and sharing them with other farmers and backyard gardeners. This is a very important way to preserve useful plants that otherwise would be lost to future generations.

Rice Growing in the Sacramento Valley

The movement toward using natural farming techniques in America is growing not only among small farmers but on some large farms as well. I am thinking in particular of the Lundberg Family Farms, which grows brown rice and is known throughout the United States. Seven years earlier I had visited the owner of this seventy-five-hundred-acre farm near Chico, California, to promote natural farming and to learn something of the way rice is grown in the vast fields of the Sacramento Valley. When he heard what I had to say, Mr. Lundberg was delighted. "This is wonderful! It *is* a revolution!" I heard later that he had taken my words to heart, had gotten rid of six of his tractors, and begun the transition to natural farming, but I did not know what happened after that.

When I saw him again on this visit, he said, "Since I met you, Fukuoka-san, I've become more open-minded. My three brothers and I are all doing natural farming now." I could see four large grain elevators standing in four different places on the farm. "Almost a hundred farmers have gathered here today to celebrate the founding of an association that aims to increase the production of naturally grown rice here in the valley, and to make it available to people around the country. I hope you will accept this medal for distinguished service from our company." He presented me with a large silver medal.

After that, a procession of cars carried us through his fields. Rice stretched as far as one could see in every direction. We stopped now and then so he could show me his trials. Finally, we left the fields behind and arrived at a cool spot shaded by a number of trees, where we had a picnic.

One thing I noticed was that these thousands of acres of rice fields were filled with barnyard grass, but Mr. Lundberg was not concerned. None of the other farmers looking at the fields seemed concerned, either. If we had been in Japan, these fields would have been considered a failure. People would have said that so many weeds growing in the fields show that the natural farming method is flawed. Then I realized how truly open-minded Mr. Lundberg had become.

Along with being impressed that he had been able to look calmly at all those weeds for seven years, I saw that the improved vitality of the soil had been the source of his success. Despite the barnyard grass, the fields produced sixteen to eighteen bushels (960 to 1,080 pounds) per quarter acre, which is about the average from conventional paddies in Japan. And he did that using large harvesting equipment while the farmers in Japan have the advantage of managing their small fields more intensively.

Until my previous visit, he had been farming organically, growing rice in each field every two or three years. The fields would then lie fallow for one year and would be planted with summer wheat or barley during the second year. After talking with me and using the natural farming method, he was able to grow rice in each field every year. Plus, because it is naturally grown rice, he got a much higher price than he would have for ordinary brown rice. It is not surprising that he has had such success.

Mr. Lundberg has also been clever about how he markets the rice, combining brown, black, and long-grained varieties into a variety of blends. He also expressed his firm resolve not to surrender to the oil companies or the capitalists who back them no matter what difficulties he may encounter.

When I saw the enthusiasm for raising new varieties of rice among the growers near Chico, in the heartland of California, I was convinced that natural farming will expand rapidly and will contribute greatly to making wholesome brown rice an everyday part of the American diet.

From Organic Farming to Natural Farming

In the future, industrial agriculture in America will probably grow even larger because of increased capital investment. At the same time, people who are inclined to use natural methods will probably progress from organic to natural farming.

The problem is, however, that the distinction between organic and natural farming is still not generally understood. Conventional farming and organic farming are actually not all that different in their approach. I consider both to be forms of scientific agriculture with modern chemical agriculture the right hand and organic farming the left.*

As far as I can see, the only way is to follow the road back to nature. I believe that by doing this, we will establish techniques that are far more appropriate than our present technology. Although this way of thinking still has various names and forms, it is based on the natural approach I describe in *The One-Straw Revolution* and *The Natural Way of Farming*.

It is fine to turn gradually from organic farming to the road that leads to nonscientific, natural farming. It is fine to set one's sights on farming that perpetuates itself sustainably, even while enjoying life on a designed farm. But these efforts should not be centered on rules and techniques. At the core there must be a sound, realistic way of seeing the world. Once the philosophy is understood, the appropriate techniques will become clear as day. Of course, the techniques will be different for different situations and conditions, but the underlying philosophy will not change. This is the most direct way to create a new agriculture that is more than just sustainable. It will provide for our needs and also heal the earth and the human spirit.

---

* Mr. Fukuoka sees little difference between modern industrial agriculture and organic farming because they both begin by addressing the same question: "How can I get nature to produce most efficiently for human beings?" The industrial farmer believes that using chemicals is the most efficient way, while the organic farmer believes that, in the long run, it is better to use organic materials. Also, both systems are based on plowing the soil, which Mr. Fukuoka does not do.

I emphasized this point when I visited students in the Agriculture Department at the University of California, Davis. The Agriculture Department at this university has played a leading role in the development of scientific agricultural in the United States. I was told that they had a reputation as simply being a research station for agribusiness.

But even there, things seemed to be changing. These days, apparently, the students were pulling the faculty toward organic and more sustainable ways of growing food. The students, as a group, were managing a farm using organic and natural farming methods, so I decided to go there, see what they were doing, and talk with them.

It was interesting that the students were growing various kinds of heirloom crops along with medicinal and ground cover plants. Still, I thought their use of green manure plants such as clover and alfalfa, which is the basis for soil improvement, was insufficient. They were giving priority to growing high-yielding crops and putting off efforts to improve the soil. I had a feeling that they were still unsure about the difference between organic farming and natural farming. If they continued in this manner, I told them they would not have a strong impact on the world.

The leader of the group, a young man originally from Ethiopia, asked the first question, and then I was showered with others. These are some of the things they asked:

"Classes in agricultural science seem to be useless. Isn't it impossible to converse with nature through books?"

"Won't the natural farming method become less beneficial to human beings as the land comes to resemble true nature more closely?"

"You talk about becoming one with nature, but how do we achieve that? Through observing nature?"

"Do you think growing ancient crops is a shortcut to returning to nature?"

"You say we shouldn't rely on human knowledge or action. Does that mean growing an improved variety of pumpkins violates the spirit of natural farming?"

"You say we should converse with the crops. Does that mean that the pumpkin may be unhappy and we can help cheer it up?"

Many novel topics and difficult questions were raised. Interesting and original answers also emerged from the group. As the students began talking excitedly in loud voices, a crowd of spectators gathered around us.

I said that the philosophical view of the universe and religious views are essentially one. The view of society and the view of life also are one. If you understand the spirit of a single flower, you understand everything. You understand that religion, philosophy, and science are one, and at the same time they are nothing at all.

It is incongruous to say, "I am a religious person. I understand the mind of God but not the mind of a pumpkin." Or, "I earn my livelihood by being a professor of philosophy, so I have no need or desire to become a farmer and grow crops."

Without understanding what it is to know things intuitively, people have sought knowledge and have become lost. Because people do not really understand what natural water is, they believe that water from the tap and the water in a river are the same.

My answers were as follows:

"The reason your classes are uninteresting is that you are listening to your professors talk about nature in classrooms illuminated by fluorescent lights. Isn't it pleasant for people just to talk to one another like we are now in the sunlight or in the cool shade of a tree?

"Ask the pumpkin whether it is happy or sad. But instead of the pumpkin, you should be asking about yourself.

"When people try to grow crops using human knowledge, they will never be anything more than farmers. If they can look at things with an empty mind as a child does, then, through the crops and their own labor, they will be able to gaze into the entire universe.

"When people are released from the idea that they are the ones who have created things and have abandoned human knowledge, nature will return to its true form. The rebirth of nature is not simply a return to the primitive, it is a return to the timeless. My method of natural farming aims at liberating the human heart. That's about as easy to understand as I can make it." When I concluded with these words, one young woman in the group blurted out, "It's so easy to understand that I didn't understand anything you said at all," and everyone burst into laughter.

When I continued, "In a word, everything is completely useless, in fact, I forgot what I just told you," another student poked fun at me saying, "That 'word' of yours was useless, too." That brought another round of laughter, and I remember thinking this was one of the most enjoyable discussions I had ever had. Finally, I told them I wanted them to keep our conversation a secret from the university professors. The students again roared with laughter.

After the talk was over, two sheepish-looking gentlemen came up to me. Once we had shaken hands, I looked at the business cards they handed me. They were professors at the university; one was the chairman of the Agriculture Department. They told me they had been impressed with what I had to say and promised to put it to good use in guiding students in the future.

Two International Conferences

The International Permaculture Conference was held in August 1986, at The Evergreen State College in Olympia, Washington, a quiet campus with a dense growth of large trees. The architecture of the main hall where the large gatherings took place was extremely original, a succession of tiers. More than seven hundred people attended the conference.

The opening remarks were delivered by one of the university professors, who was a Native American. He wore a feathered headdress, and his stately ceremonial attire was breathtaking. I was very impressed with his address in which he cited ancient American Indian legends with regard to the relationship his people have with nature. It reminded me that I truly was in America.

On the first day there were introductory remarks given by people from many countries. The main event consisted of talks given by Bill Mollison, the co-creator of permaculture, from Australia; Wes Jackson, the founder of The Land Institute in Salina, Kansas; and me.

Mollison's permaculture is a no-tillage system that uses perennial plants and trees to create abundant farms and resilient human communities designed after natural landscapes. The farms are meant to perpetuate themselves indefinitely without bringing in materials from the outside. It is based on organic agriculture and appears to have quite a following in Australia, the United States, and around the world.

Jackson aims to develop farming methods that use as little fossil fuel as possible. If we do not limit the use of fossil fuel, he believes, there will be no future for agriculture. He is working to develop native perennial grasses into food crops that will eliminate the need for plowing. It appears that, while Jackson fundamentally accepts the scientific approach, he is searching for the beginnings of a new agriculture.

I was introduced by the moderator as "an advocate of natural farming, which is founded on the philosophy of *mu* (nothingness) and disavows the value of science."

The idea was to find common ground from our three distinctive viewpoints and set a new, better course for agriculture in the future. On the following day, the three of us held a panel discussion. We were lined up together on the stage, and the discussion took place in a question-and-answer format, debate-style.

The questions had been given to us the day before, and the discussion was proceeding smoothly in accordance with the general outline, until it unexpectedly dissolved into a slapstick comedy that had everyone roaring. Mollison spoke English with a strong Australian accent, so Jackson teased him, saying he could not understand a word Mollison was saying. To top it off, I had three translators, and whenever I said something, they would give three different interpretations.

Someone in the audience joked that they had no idea what Fukuoka was really saying. The audience became aware of how difficult it is to penetrate the heart of Oriental languages and expressions. People were both perplexed and intrigued, and the exchange of strange questions and even stranger answers continued amid gales of laughter.

Finally I drew a picture of Don Quixote's donkey. On its back were a blind Bill and a deaf Wes both riding backward, and me hanging desperately on to the donkey's swishing tail. The three Don Quixotes, hoping to return to nature, were trying to stop the donkey from rushing wildly toward the brink of disaster, but it seemed hopeless. Someone asked what was going to happen, so I drew President Reagan sitting frontward on the donkey's back dangling a carrot in front of the donkey's nose. When I asked, "What do suppose the carrot is?" someone correctly answered, "Money."*

The next conference I attended was one week later at the University of California, Santa Cruz. The university campus was built on an extensive site following a new concept. Once you enter the gate, you find yourself amid a forest of oak, maple, and redwood trees. In order to preserve the forest, the buildings were constructed so that they seem to appear and disappear among the trees, and the distance between buildings is often too far to walk. Most of the students ride bicycles.

My talk was last on the agenda, so I was allotted plenty of time. The auditorium was filled to capacity. Half of the people there were adherents of the nature

---

* This drawing is reproduced at the beginning of this chapter.

movement from around the world, while the other half were scientists connected with the university. I was told to assume that many of them were supporters of scientific farming. The air was filled with tension.

I interspersed jokes with my talk, and I think the audience enjoyed listening to me. But when the translator began to read my final statement, which disavowed the value of science, the hall suddenly fell silent. I had no idea what had happened and was rather concerned. The moderator then called for comments. Two people who had raised their hands were given an opportunity to speak.

The first person was from India. He said that my ideas were similar to Gandhi's, and that the ancient Indian texts confirm that no-tillage agriculture was practiced there long, long ago. He concluded by voicing approval for my farming methods and ideas.

The man who stood next was a professor of both religion and philosophy. He had disagreed with me the day before at one of the seminars. He was critical of the philosophy of Socrates while I defended the Greek thinker by saying that even though I rejected all the Western philosophers from Descartes on down, I could not deny the truth of even a single syllable of what Socrates said. We reached no conclusion and agreed to talk again the following day.

Of course, I expected him to counter my argument, but instead he faced the audience and launched into a five-minute speech. "Mr. Fukuoka has explained how, in the course of Western philosophy, Descartes, Locke, Kant, Hegel, and others established the foundation of modern science. Moreover, he negated the basic principles established by these philosophers and was successful in substantiating his argument. I found that quite startling. We cannot avoid recognizing that modern scientific agriculture has been based on flawed principles. I welcome Mr. Fukuoka's philosophy and farming methods as a new approach that will help create a better future for the world."

When his speech was finished, there was thunderous applause. I was overwhelmed with emotion, thinking that there had indeed been a purpose in my coming to the United States. I felt as if a heavy burden had been lifted from my shoulders.

Japanese Cedars at the Zen Center

After visiting some places on the East Coast, and before I started my journey home, I stopped at the Zen Center's Green Gulch Farm outside San Francisco, a place I had visited before seven years earlier. The farm is surrounded by bare, savanna-like hills, but considering that there is a redwood forest in a nearby national park, it seems likely that in ancient times this area was forested.

My guide on my previous visit had been Harry Roberts.[*] Mr. Roberts was a Native American who

---

[*] Born in 1906, Harry Roberts was a member of the Yurok tribe. He went to school in the San Francisco Bay Area, but spent summers with the Yuroks on the Klamath River in Northern California. Mr. Roberts was trained for twenty years to be the "high man" of his tribe, the one who is entrusted with carrying on the lineage of his people. During his eventful life he was a blacksmith, cook, cowboy, horticulturist,

was helping the students with their farming. This time, unfortunately, he was no longer there. During my previous visit, Mr. Roberts and I had walked together through the redwoods at Muir Woods. We saw huge trees 200 to 250 feet tall with sword ferns and huckleberries growing over a large area. I told him that the ecosystem of mixed trees and understory plants there closely resembled the virgin forests of Japan. We concurred that this fact gave us a clue for revegetating California forestlands. When we parted, I had told him, "You are the guardian deity of the American forests. You are a giant both in body and spirit." He replied, "You are small in stature, but to me you are a giant of the Orient," and everyone had a good laugh.

Redwood trees grow quickly to a large size, but their roots are shallow, so they easily fall over unless they are supported by the root systems of their neighbors. This is especially true now since so much of the topsoil has eroded in their natural range. In contrast, the Yaku Island and Yanase *Cryptomeria*, or Japanese cedars, which are related to redwoods, send down deep roots that can penetrate the subsoil. I promised to send him seeds of those varieties. After returning to Japan I did send him a handful of seeds, and in return he sent me a cup made from the wood of a redwood tree.

I had heard that he carefully planted those *Cryptomeria* seeds, and when I visited the Zen Center

---

naturalist, spiritual adviser, and more. In the 1950s, he planted the first *Metasequoia* trees in California. These trees can be found in San Francisco's Golden Gate Park, in the University of California's Botanical Garden in Berkeley, and in parks and yards throughout Northern California.

this time and the students came to greet me, the first thing they did was show me a photograph of Mr. Roberts.

The photograph showed this great elder on his deathbed, surrounded by his disciples. He had raised himself into a sitting position and was planting seeds in a seedling flat. They told me he had said, "These seeds are Fukuoka's spirit. Sow them carefully, and when the seedlings have developed, plant them in those three valleys over there." Shortly after giving these instructions, he passed away.

When I thought about how he, with his large body that resembled the images of the recumbent Sakyamuni Buddha entering nirvana, had cared so much about the seeds I had sent and had been so deeply concerned about regreening the forests and deserts, a lump rose in my throat, and I was unable to speak.

Ten or twenty people took me to see the place where he had told them to plant the seedlings. Many *Cryptomeria* saplings were growing there; some of them were already six feet tall. Iron stakes had been driven into the ground around each tree, and wire fences were surrounding them. I was told that this was to protect them from deer and other wild animals. I realized just how much trouble these young people had gone to.

Someone said, "Our teacher must be glad you came to see this place. He is resting on that hillside over there." When I looked across the valley in the direction they were pointing, I could see a spot of about twenty square feet where some stones had been crudely piled up. It was similar to grave sites I had seen in the desert of Somalia.

One person said, "Our teacher must be calling out to you, saying 'Let's plant seeds in the desert together.'"

Jokingly I answered, "It looks like a comfortable place. It might not be bad to lie down there with him." But as soon as I had said this, I burst into an uncontrollable flood of tears.

Yes, it was true. The man lying there had been a sower of seeds in the desert. When I thought how he might be the only person I would ever meet who understood me, who would live with me and die with me, I stood rooted to the spot, heedless of the tears pouring down my cheeks.

I have no idea why I was crying so. I had not even shed tears when my own mother and father died. This was only the second time I had cried in fifty years. The first time was on my earlier visit, when I spoke at the summer camp held in the midst of the wonderful forest at French Meadows in the Sierra Nevada. I was recalling the spring of my twenty-fifth year, when I underwent my change of heart, and had just asked the group, "What is true nature?" Suddenly words failed me, and tears flowed from my eyes. I had to stop my talk for several minutes. This situation was completely different, but I have the feeling they were the same tears.

Mr. Roberts was no longer there. Neither his body nor his spirit was in this world anymore. It was because I knew that his spirit was not even drifting about in some other world that I was able to cry. I sensed that those tears flowed from a place that transcended life and death and that, in fact, I had been bathed in tears of ecstasy.

The people of the Zen Center must have had the same feeling. They left me alone while I gazed up at the blue California sky. Then, finally, we began to talk happily again as we headed back to the Zen Center.

APPENDIX A

Creating a Natural Farm in Temperate and Subtropical Zones

When starting a natural farm, the first question you might ask is where it should be located and how you will choose the place to live on it.

You could go into a mountain forest and live in splendid isolation, if you like that sort of thing, but usually it is better to make a farm in the mountain foothills. The climate is better in an area that is slightly elevated from the valley floor; there is also less chance of flooding. It is easy to obtain firewood, grow vegetables, and find the other materials needed for satisfying the necessities of food, clothing, and shelter. If there is a river nearby, crops will be easier to grow, and you can establish a satisfying life for yourself.

No matter what the land is like, you can grow crops if you make the effort, but it is easier if the place is rich with the bounties of nature. It is ideal to find a place where large trees grow densely on the hills, the soil is deep and black or dark brown in color, and clean water is available. The natural farm should include not only fields but also the surrounding hills and forests. A good environment and fine scenery are wonderful elements for living an enjoyable life, both materially and spiritually.

Natural Protective Forests

The foundation for achieving success is to build deep, fertile soil. Here are some methods for achieving this.

1. Burying coarse organic matter such as decaying tree trunks and branches in the ground is one way, but this requires a great deal of labor. In general it is better to let plants do the work for you.

2. Plant trees, shrubs, grasses, and legumes with extensive root systems. These roots will soften the soil and bring nutrients from the lower levels back up to the surface, gradually improving it.

3. Cause rainwater to flow over the farm from the wooded slopes above. This water will carry nutrients from the humus-rich forest soil. The essential thing is to maintain a continuous supply of organic material that is produced on the farm itself. This forms the basis of fertility for the entire system.

You can improve the protective forests on the slopes above the farm, but if there are no forested hills, new woods or bamboo thickets should be created. When creating or improving the protective forest, you should plant a mixture of trees, shrubs, and ground cover plants that are useful for several purposes. For example, they may be useful as a source of fuel and building materials, for providing habitat for birds and insects, and for producing food for farm animals, wildlife, and people. It is also good to include plants that have medicinal benefits, attract insects, and improve the soil. Often one plant fulfills many of these functions. The idea is to have all of the elements on the farm working together as one.

Raising a Protective Forest

The soil at the summit of a hill or on the ridgeline of mountains is often thin and dry leaving the hill bare at the top. In places like this, we should first grow trailing plants like *Ixeris debilis* and kudzu (*Pueraria lobata*)* to halt the erosion of soil and then plant or sow the seeds of pines and Sawara cypress (*Chamaecyparis pisifera*), turning it into an evergreen forest. At first, fast-growing grasses such as eulalia (*Miscanthus sinensis*) and cogon grass (*Imperata cylindrica*), ferns such as bracken (*Pteridium aquilinum*) and scrambling fern (*Diplopterygium* spp.), and bushes such as bush clover (*Lespedeza* spp.), *Eurya japonica*, and cypress will grow densely. Gradually, as the soil improves, these plants will

---

* Kudzu is a fast-growing vine native to Japan and southern China. It is very useful in rehabilitating poor or damaged soil, and its roots, leaves, and flowers are widely used in Asian cooking. Since its introduction to the United States, it has become rampant in many areas, especially in the Southeast. It is not rampant in Japan. Plants mentioned in this section are representative of the plants Mr. Fukuoka used for various purposes while developing his natural farm in Japan. Other plants that perform the same functions, but are better suited to local conditions, should be substituted as appropriate.

be replaced by a succession of other plants. Eventually, miscellaneous trees will begin to grow here and there from seeds deposited by the plants themselves, and by animals, birds, and the wind.

On the side of the hill, it is good to plant evergreens such as hinoki cypress (*Chamaecyparis obtusa*) and camphor, along with a mixture of other trees such as Chinese nettle tree (*Celtis sinensis*), Japanese zelkova (*Zelkova serrata*), paulownia (*Paulownia tomentosa*), cherry, maple, and eucalyptus. The foothills and the valley floor usually have deeper, more fertile soil, so you can plant trees such as walnuts and ginkgo among evergreens such as Japanese cedar (*Cryptomeria japonica*) and oak trees.

Bamboo groves are also useful in protective forests. Bamboo grows to full height from shoots in a single year. Its volume of growth is greater than that of the typical woody tree, so it is quite valuable as a source of coarse organic material when buried.*

The shoots of Moso bamboo (*Phyllostachys edulis*) and other bamboos are edible. When dried, the stalks are light and easy to carry. Bamboo decomposes slowly when buried, so it is very effective for holding water and air in the soil. The organic matter produced as a result of its decomposition is wonderful for improving the structure of the soil.

---

* Bamboo is a widely adapted, fast-growing grass that can easily be grown and managed sustainably. It has countless uses in the broad categories of fiber for clothing and paper, household utensils, firewood, building, and food. There are two main types of bamboo, clumping and running. Care should be taken to plant the right variety of bamboo, as running bamboo can sometimes become difficult to control.

Windbreaks

Trees planted as windbreaks are valuable not only for preventing wind damage but also for maintaining the fertility of the soil, filtering water runoff, and generally improving the environment for all creatures. Varieties that mature rapidly are Japanese cedar, hinoki cypress, acacia, and camphor. Camellia, parasol fir (*Firmiana simplex*), strawberry tree (*Arbutus unedo*), and star anise (*Illicium verum*), while slow growing, are also exceedingly beneficial. Depending on conditions, you may also be able to use trees such as evergreen oak, Japanese cleyera (*Ternstroemia japonica*), and holly.

Creating an Orchard

To create a natural orchard, take the same approach you would in planting trees in the forest. Cut trees in stages, taking nothing from the fields—that means the large trunks, branches, and leaves are all left to decay in their own time. These trimmings can be piled up along the contour lines of slopes to serve as improvised terraces.

The idea is to create a mixed orchard without clearing the land, only thinning. These days when land is cleared for an orchard, it is usually done with a bulldozer. The uneven surfaces of the slopes are flattened and roads are created to facilitate mechanized management.

Using machinery to create roads in the orchard makes it easier to apply fertilizer and other agricultural chemicals, but this is not necessary with the natural method. The only heavy labor is picking the fruit, and that is done by hand. I believe that success is actually more likely if you have little money when beginning the orchard. That way you introduce neither machines nor capital.

The leaves and branches of trees that were cleared and left on the ground, along with the tree roots, will slowly decay, becoming a long-term source of organic fertilizer. They will provide nutrients for the fruit trees for about as long as it takes the young orchard trees to grow to the size of the woodland trees that were thinned.

Furthermore, the organic matter provides a welcoming environment for ground cover to grow. The ground cover helps to suppress weeds, prevents soil erosion, stimulates the soil microorganisms, and improves the soil's structure and overall vigor.

Just like trees planted in a forest, the fruit trees are best planted along the contour of the slope. If possible, you should dig holes for planting, add coarse organic matter, and then plant the trees in a raised mound on top of that. Trees that are planted and grown in a natural way live longer and are more resistant to insects and disease than those grown with extra fertilizer and agricultural chemicals.

One of the problems with using a bulldozer to create an orchard is that when you flatten the land, you scrape off the topsoil. The topsoil contains the majority of the organic matter that has accumulated for many years. A farm cleared by a bulldozer and left untouched for ten years dries out and loses soil to erosion; its economic value is greatly reduced. Another problem with conventional clearing is that the trees are clear-cut and burned. At a single stroke, the fertility of the soil is diminished for decades.

In about twelve inches of topsoil, there are enough nutrients to sustain an orchard for ten years without adding fertilizer. If you have three feet of rich earth, the orchard can be sustained for approximately thirty years. If we can retain and maintain the richness of natural forest soil by using a soil-building combination of plants, including nitrogen-fixing plants such as white clover, beans, and vetch, then no-fertilizer cultivation is possible indefinitely.

Creating "Fields" in the Orchard

Usually a "field" means a place devoted exclusively to growing crops, but the spaces between the fruit trees in an orchard can also be considered a field. The system and methods of cultivation, however, differ greatly depending on whether the orchard or conventional flat fields predominate. Making a field where fruit trees are the primary crop and grains and vegetables are grown between the trees is almost exactly the same as creating an orchard for natural farming. It is not necessary to clear the land, and there is no need to carefully prepare the ground or bury coarse organic matter.

In the first stage of making the orchard field, we need to control weeds and bring the soil to maturity. It is good to start by sowing the crops among green manure plants and buckwheat during the first summer, and radishes and mustard during the first winter. In the next year you can plant strong twining plants that reproduce well without fertilizer, such as adzuki beans and cowpeas (black-eyed peas) in summer, and hairy vetch in winter. These twining plants, however, can cover up vegetables and fruit tree seedlings, so they must be tended from time to time. As the field matures, you can grow a wide variety of other crops.

Creating a Conventional Field

Most lowland field crops are annuals that are produced in just a few months to half a year. These are the typical garden vegetables. The ones that reach about two or three feet in height, like tomatoes, eggplant, and peppers, have shallow roots and are a bit temperamental because of their long history of hybridization. Because the time between sowing and harvesting fast low-growing crops such as radish, lettuce, and garden turnips is short, and several crops are typically grown in one year, the surface of the ground is exposed to the elements for much of the growing season. We must accept the fact that some soil loss will occur in these fields because of rain, and that the soil will lose some of its vitality during droughts and cold weather. These problems can be minimized, however, by keeping the soil covered with mulch and by growing a continuous ground cover.

When creating the field, the most important concern is preventing soil erosion. If the land is sloped, it should be terraced to make the surface level or close to level. This can be accomplished by building up earthen banks or making stone walls, then creating the terraced fields. The success or failure of this work depends on knowing the type of soil so the banks will not crumble.

For the stone walls, it is ideal to use the stones dug up from the field or hillside itself.

Creating Paddy Fields

Of course, it is easy to make paddy fields by clearing the plains with bulldozers, carrying away the vegetation, and leveling the ground. This also makes it possible to increase the size of the fields, making mechanized agricultural more convenient.

There are many disadvantages to this method, however. (1) The topsoil of the paddy is of uneven depth, resulting in uneven growth of crops. (2) Because the large, heavy machinery puts pressure on the soil, the soil becomes compacted. Groundwater will collect and stagnate. This creates anaerobic conditions, which cause the roots to decay, leading to damage from disease and insects. (3) When a field is created by using heavy machinery, the ridges around the fields become hard, the microorganisms in the soil change or die out, and the soil atrophies.

Trees are the guardians of the soil. Even in flooded paddies, growing large and small trees on mounds right in the fields themselves is an excellent idea. The paddies near Sukhothai, Thailand, are filled with such trees. Those fields are among the finest examples of the natural faming method for growing paddy rice anywhere in the world since they join the farmers with a diversity of plants and animals—including draft animals, fish, and amphibians—into a harmonious whole.

It is unnecessary to go to great lengths to grow rice in paddy fields, since it is quite possible to grow paddy rice in dry fields watered only by the rain. I have shown this in my fields in Japan.* The main reason people grow rice in a flooded field is to control weeds. I take care of weeds by not plowing, spreading straw mulch, and growing white clover continuously on the surface of the soil. I get better results that way without doing much at all, and the soil improves with each passing year.

---

* This is possible on Mr. Fukuoka's farm partly because he gets reliable rainfall throughout the growing season. Rice can also be grown successfully in dry, upland areas with little or no irrigation, but the yields are lower.

APPENDIX B

Making Clay Seed Pellets for Use in Revegetation

Purpose

THE CLAY SEED PELLET was conceived and developed for direct seeding rice, barley, and vegetables in conjunction with the no-till method. It has since come into wide use, and is particularly well suited for aerial seeding for the purpose of revegetating large areas of desert at one time.[*]

Materials

1. Seeds of more than one hundred varieties (trees, fruit trees, shrubs, vegetables, grains, useful fungi). 10 percent of combined weight.

---

[*] Farmers around the world have been encasing seeds in clay pellets for years. Mr. Fukuoka did not invent the technique; he has simply revived it. As a result, clay pellet seeding is again being used effectively on small-scale farms, for large rehabilitation projects, and in urban areas by guerrilla gardeners to seed particularly dry, compacted sites.

2. Fine powdered clay such as that used for fired bricks or porcelain. In general, this should make up five times the weight of the seeds, but the amount of seeds should be taken into consideration. 50 percent of combined weight.

3. Bittern—the liquid remaining after salt has been removed from the brine obtained by boiling and concentrating seawater or from natural brackish water (such as the water found in the Dead Sea). 10 to 15 percent of combined weight, with seaweed paste for binding making up 5 percent of combined weight.

4. Slaked lime—10 percent of combined weight.

5. Medicinal herbs: derris (root), powdered fruits and leaves of Japanese star anise (*Illicium anisatum*), Japanese andromeda (*Pieris* spp.), Japanese lacquer tree (*Rhus verniciflua*), Japanese bead tree (*Melia azedarach*). 10 percent of combined weight.

6. Water—5 to 10 percent of combined weight.

Aerial Seeding Method

The seeds necessary for revegetation of the desert will be mixed in clay pellets and broadcast from airplanes or by hand to revegetate large areas at one stroke.

Method of Production

When producing pellets in large quantities, a typical concrete mixer (with inner blades removed) is useful.

1. Put fungi and seeds into the mixer and mix well to spread the fungi about (inner layer).

2. Next, alternately add the clay powder with a water mist in the mixer as it is rotating, to create a layer enclosing the seeds and fungi (middle layer).

3. Then, when you alternately add and spray the bittern, seaweed paste solution, clay powder, and slaked lime into the mixer as it is rotating, a round seed pellet usually about a quarter inch to a half inch in diameter will form (outer layer).

Properties

1. The seeds enclosed in the layers of clay will achieve satisfactory germination and growth with the aid of the useful fungi.

2. By kneading the clay together with the bittern and seaweed paste, its molecules are rearranged, so the pellets become stable, light, and hard. They will not only withstand the fall to earth following aerial seeding, but can also adjust to changes in dampness and dryness related to rainfall, becoming shrunken and solid. Thus, they seldom crumble or break, and

the seeds are protected from damage by birds or animals until they germinate.

3. Many insects are repelled by the bitterness of the herbs and the bittern mixed into the outer layer, so the seeds escape being eaten. In deserts and savannas this technique helps prevent damage by mice, goats, and, in particular, strong insects such as red ants. Even damage by birds can be prevented simply by enclosing the seeds in pellets. The method described here not only ensures safe germination of seeds in desert areas without the use of toxic substances, but also makes it possible to indiscriminately broadcast the seeds over a wide area.

4. The plants on earth exist in intimate connection with other plants, animals, and microorganisms, and none can develop and flourish alone. In desert regions, in particular, microorganisms are necessary as well as a variety of plants.

5. Derris root (used against beetles), Japanese star anise (goats), Japanese andromeda (cows), Japanese bead tree (small insects), sumac, and so on, will protect seeds in the desert, before and after germination.

In a region that is completely desert, it is a good idea to mix fertile jungle or forest soil with the clay. This soil is a rich source of soil microorganisms, seeds, and spores, and is of great value when added to the pellets.

If the pellets will be broadcast from airplanes, the pellets may break on contact with the ground, so it is good to coat them with seaweed paste, if available.

Thus, even in vast desert areas, where conditions for germination are poor, revegetation can be achieved simply by sowing the seeds, without concern about time or place. Successful results have already been achieved in Africa, the United States, India, Greece, and the Philippines.*

---

* Even without using all of the additives Mr. Fukuoka recommends in this section for broad scale sowing in the desert, simply encasing seeds with clay to form pellets will greatly aid germination in most places.

APPENDIX C

Producing an All-Around Natural Culture Medium

THE ALL-AROUND NATURAL culture medium (matsutake fungus culture medium) for use with clay pellets is comprised of bacterial and fungal microorganisms, especially mycorrhizal fungi (matsutake), eumycetes (fungi), actinomycetes (fungi), and bacilli (bacteria).

This culture medium is made by combining leaf mold and rice bran, and adding to them an infusion derived from boiling a fixed amount of the tubers and/or stems of plants of the Convolvulaceae family (morning glory) and/or Dioscoreaceae (yam) family, a fixed amount of plants of the Cruciferae (mustard, cabbage) family, and a fixed amount of plants of the Liliaceae (onion, garlic) family.

Although the pure isolation of the matsutake fungus was achieved long ago, research on the artificial culture of matsutake, especially through a pure cultured fungus, has proceeded very slowly. One reason for this is that isolated culture of the hyphae is very difficult to achieve. Another problem is that the hyphae grow extremely slowly, and it is difficult to obtain a large amount of them. (No other fungus is as difficult to culture as matsutake.)

The matsutake fungus culture medium described here overcomes these drawbacks. After experimenting with various media, I was able to obtain a large amount of hyphae quite easily with the natural culture medium. The rate of growth and development of the hyphae in this medium is approximately ten times faster than in the Hamada medium, which has been used most often in the past. Moreover, the medium produced a dense, vigorous growth of the hyphae.

The aforementioned natural culture medium can be produced using the following materials:

1. Leaf mold from pine forests or mixed woods (20 to 50 percent of weight)

2. Rice bran (20 to 50 percent of weight)

3. Vegetables
    a. Tubers and stems from members of the Convolvulaceae family such as sweet potatoes and members of the Dioscoreaceae family such as yams (10 to 20 percent of weight)
    b. Crucifers such as daikon and mustard (10 to 20 percent of weight)
    c. Members of the Liliaceae family such as onions, wild onions (ramps), and wild garlic (10 to 20 percent of weight)

Crush and mix the leaf mold and rice bran to make up 20 percent of the total weight of the culture medium. Then add to this mixture an infusion made from boiling material from the three vegetable groups, each making up 20 percent of the weight. You can make either a liquid medium, with the addition of 60 to 80 percent water, or a solid medium, with the inclusion of agar. In order to artificially culture the matsutake fungus using this medium, first put the medium in jars or plastic bags and sterilize them with steam, then inoculate them with the spores, and keep them in a hothouse at a constant 16 to 23 degrees Celsius (61 to 73 degrees Fahrenheit). After about one month the hyphae will have spread throughout the containers. Then, if you move them to a location maintained at 19 degrees C (66 degrees F) or colder, fruiting bodies will develop after about four months.

Why is this natural medium effective? The nutritional sources of the matsutake are subtle and complex. Even if we put together various chemical compounds and add vitamins and hormones, we cannot obtain a large quantity of hyphae easily, and it is almost impossible to see the development of fruiting bodies.

The natural medium, on the other hand, may appear at first to be casually made, but it exhibits an organic, synergistic effect in which the various elements work together. It appears to enhance the nourishment of the hyphae and the formation of the fruiting bodies. Vegetative propagation of the hyphae is possible to a certain extent just with the leaf mold and rice bran, but it appears that the vitamins and hormones, especially the growth hormones, contained within the vegetables play a large part in the formation of the fruiting bodies.

This natural culture medium is suitable not only for cultivating matsutake fungi, but for cultivating microorganisms in general.

About the Author

MASANOBU FUKUOKA (1913–2008) was a farmer and philosopher who was born and raised on the Japanese island of Shikoku. He studied plant pathology and spent several years working as a customs inspector in Yokohama. While working there, at the age of twenty-five, he had an inspiration that changed his life. He decided to quit his job, return to his home village, and put his ideas into practice by applying them to agriculture.

In 1975 he wrote *The One-Straw Revolution*, a best-selling book that described his life's journey, his philosophy, and farming techniques. This book has been translated into more than twenty-five languages and has helped make Mr. Fukuoka a leader in the worldwide sustainable agriculture movement. He continued farming until shortly before his death in 2008, at the age of ninety-five.

About the Editor

LARRY KORN is an American who lived and worked with Mr. Fukuoka on his farm for more than two years in the 1970s. He is the translator and editor of Mr. Fukuoka's first English-language book, *The One-Straw Revolution*, and accompanied Mr. Fukuoka on his visits to the United States in 1979 and 1986. Korn studied Chinese history, soil science, and plant nutrition at the University of California, Berkeley. He currently lives in Ashland, Oregon, giving workshops and classes on natural farming, permaculture, local food production, and building resilient, sustainable communities.

Chelsea Green Publishing is committed to preserving ancient forests and natural resources. We elected to print this title on 30-percent postconsumer recycled paper, processed chlorine-free. As a result, for this printing, we have saved:

**14 Trees (40' tall and 6-8" diameter)
6 Million BTUs of Total Energy
1,405 Pounds of Greenhouse Gases
6,335 Gallons of Wastewater
402 Pounds of Solid Waste**

Chelsea Green Publishing made this paper choice because we and our printer, Thomson-Shore, Inc., are members of the Green Press Initiative, a nonprofit program dedicated to supporting authors, publishers, and suppliers in their efforts to reduce their use of fiber obtained from endangered forests. For more information, visit: www.greenpressinitiative.org.

Environmental impact estimates were made using the Environmental Defense Paper Calculator. For more information visit: www.papercalculator.org.

the politics and practice of sustainable living
CHELSEA GREEN PUBLISHING

Chelsea Green Publishing sees books as tools for effecting cultural change and seeks to empower citizens to participate in reclaiming our global commons and become its impassioned stewards. If you enjoyed reading *Sowing Seeds in the Desert*, please consider these other great books related to Gardening and Agriculture.

GAIA'S GARDEN
A Guide to Home-Scale Permaculture,
Second Edition
TOBY HEMENWAY
9781603580298
Paperback • $29.95

THE RESILIENT GARDENER
Food Production and
Self-Reliance in Uncertain Times
CAROL DEPPE
9781603580311
Paperback • $29.95

THE SEED UNDERGROUND
A Growing Revolution to Save Food
JANISSE RAY
9781603583060
Paperback • $17.95

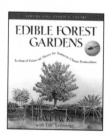

EDIBLE FOREST GARDENS (two-volume set)
Volume I: Ecological Vision and Theory
for Temperate Climate Permaculture
Volume II Ecological Design and Practice
for Temperate Climate Permaculture
DAVE JACKE WITH ERIC TOENSMEIER
9781890132606
Hardcover • $150.00

For more information or to request a catalog, visit **www.chelseagreen.com** or call toll-free **(800) 639-4099**.

the politics and practice of sustainable living

CHELSEA GREEN PUBLISHING

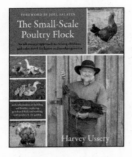

THE SMALL-SCALE POULTRY FLOCK
An All-Natural Approach to Raising Chickens and Other Fowl for Home and Market Growers
HARVEY USSERY
9781603582902
Paperback • $39.95

THE CHINESE MEDICINAL HERB FARM
A Cultivator's Guide to Small-Scale Organic Herb Production
PEG SCHAFER
9781603583305
Paperback • $34.95

THE HOLISTIC ORCHARD
Tree Fruits and Berries the Biological Way
MICHAEL PHILLIPS
9781933392134
Paperback • $39.95

SEPP HOLZER'S PERMACULTURE
A Practical Guide to Small-Scale, Integrative Farming and Gardening
SEPP HOLZER
9781603583701
Paperback • $29.95

For more information or to request a catalog, visit **www.chelseagreen.com** or call toll-free **(800) 639-4099**.